The Past, Present, and Future of Integrated History and Philosophy of Science

Integrated History and Philosophy of Science (iHPS) is commonly understood as the study of science from a combined historical and philosophical perspective. Yet, since its gradual formation as a research field, the question of how to suitably integrate both perspectives remains open. This volume presents cutting edge research from junior iHPS scholars, and in doing so provides a snapshot of current developments within the field, explores the connection between iHPS and other academic disciplines, and demonstrates some of the topics that are attracting the attention of scholars who will help define the future of iHPS.

Emily Herring, Kevin Matthew Jones, Konstantin S. Kiprijanov, and Laura M. Sellers are postgraduate researchers based at the Centre for the History and Philosophy of Science at the University of Leeds.

History and Philosophy of Technoscience

Series Editor: Alfred Nordmann

The Past, Present, and Future of Integrated History and Philosophy of Science

Edited by
Emily Herring, Kevin Matthew Jones,
Konstantin S. Kiprijanov and Laura M.
Sellers

Routledge
Taylor & Francis Group

LONDON AND NEW YORK

First published 2019
by Routledge
2 Park Square, Milton Park, Abingdon, Oxon OX14 4RN

and by Routledge
605 Third Avenue, New York, NY 10017

First issued in paperback 2021

Routledge is an imprint of the Taylor & Francis Group, an informa business

British Library Cataloguing-in-Publication Data
A catalogue record for this book is available from the British Library

Library of Congress Cataloging-in-Publication Data
A catalog record has been requested for this book

ISBN 13: 978-0-367-78638-0 (pbk)
ISBN 13: 978-0-8153-7985-0 (hbk)

Typeset in Times New Roman
by Taylor & Francis Books

Contents

Figures

Contributors

Alex Aylward is currently working towards his PhD in the Centre for History and Philosophy of Science at the University of Leeds, UK. His research focuses upon the influential British statistician and geneticist R. A. Fisher (1890–1962), and in particular the writing and reception of his *The Genetical Theory of Natural Selection* (1930), considered by some to be the most important book on evolution since Charles Darwin's *Origin*. Alex's interests lie mainly in the history and philosophy of the biological sciences, but also extend to historiography, and the relations between history of science and philosophy of science.

Klodian Coko's research focuses on the historical emergence and development of scientific methods. He is currently a postdoctoral associate in the Rotman Institute of Philosophy, at the University of Western Ontario. He has previously been a postdoctoral fellow in the Edelstein Center for the History and Philosophy of Science, Technology and Medicine, at the Hebrew University of Jerusalem (2016–2017), and a predoctoral fellow at the Max Planck Institute for the History of Science (2014). In December 2015, he was awarded a PhD in History and Philosophy of Science from Indiana University Bloomington.

Claudia Cristalli is currently doing a PhD at UCL, London. Her thesis, 'The Philosophical Psychology of Charles S. Peirce', addresses a neglected aspect of Peirce's engagement with science and explores how Peirce's pragmatist philosophy shaped his inquiry in cognitive psychology. She recently published 'Experimental Psychology and the Practice of Logic' in the *European Journal of Pragmatism and American Philosophy*. Her interests include philosophy of science, philosophy of history, philosophy of science in practice, philosophy of the social sciences, feminist epistemology. Outside academia, Claudia's commitment to philosophy includes facilitating philosophical discussion and presenting philosophical topics at Stuart Low Trust Philosophy Forum.

Joe Dewhurst is currently a Postdoctoral Fellow at the Munich Center for Mathematical Philosophy, where he is working on developing a moderately perspectival account of mechanistic functions in biology and cognitive science. He was awarded his PhD in 2017 from the University of Edinburgh, where his doctoral research looked at the relationship between common-sense intuitions and scientific theories in contemporary cognitive science. His other research interests include the philosophy of computation, philosophy of mind and cognition more generally, and the history of cognitive science.

Andrea Gambarotto completed a Cotutelle Doctoral Degree Program under the joint supervision of the Istituto Italiano di Scienze Umane (Florence) and the Institut d'histoire et de philosophie des sciences et des techniques (Paris). He is currently a post-doctoral researcher at the Université Catholique de Louvain, where he is developing a project on Hegel's 'philosophy of biology'.

Emily Herring studied philosophy at the Sorbonne and is now completing a PhD at the University of Leeds on the reception of French philosopher Henri Bergson's theories among British biologists.

Jon Hodge is Senior Fellow in History and Philosophy of Science at the University of Leeds. His books include *Origins and Species* (1991) and two volumes of papers, *Before and After Darwin: Origins, Species, Cosmogonies and Ontologies* (2008) and *Darwin Studies: A Theorist and His Theories in Their Contexts* (2008). Jon has also edited the *Routledge Companion to the History of Modern Science* (1996) with Geoffrey Cantor, John Christie and Roger Olby, and the *Cambridge Companion to Darwin* with Greg Radick (2nd edition, 2009).

Kevin Matthew Jones is a postgraduate researcher at the University of Leeds, researching topics within the history and philosophy of psychiatry, in particular attempts made to standardize psychiatric classification in the United Kingdom c.1800–1950. He has published on the history of psychiatric and psychological concepts, popular representations of psychiatric concepts, and the history of psychology. His most recent paper looked at intersections between psychiatry and psychology within the British Psychological Society's section on Medical Psychology, and the implications of its disappearance at the end of the 1950s. He has also written about music, and recently contributed a chapter to *Fight Your Own War: Power Electronics and Noise Culture*.

Konstantin S. Kiprijanov is a postgraduate researcher at the Centre for the History and Philosophy of Science, University of Leeds. His research focuses on the history and philosophy of chemistry, broadly construed. Konstantin's PhD thesis investigates the making of the modern chemical notation from an interdisciplinary and transnational perspective. Konstantin also has strong research interests in the epistemology of visual representations, Soviet science, and the role of communication practices in the making of scientific knowledge. His most recent article investigates the history of the so-called Belousov-Zhabotinsky reaction and its contribution to the nascent research field of nonlinear science.

Wonyong Park is a graduate student at the Department of Physics Education, Seoul National University, Korea, where he earned a bachelor's degree and was trained as a physics teacher. His research focuses on the intersection of history, philosophy, sociology of science and education theory, modern physics curriculum and science teacher education. His recent work on the implications of Goethe's natural philosophy for practical science education was published in the *Science & Education* journal. During his master's programme, he has received scholarships and research grants from Seoul National University and NARST International Committee.

Eugenio Petrovich is Research Fellow at the Department of Political Economy and Statistics of the University of Siena (Italy), where he is working on a research project on the impact of research evaluation exercises on the publication and citation habits of scientists. Previously, he was a PhD candidate at the Doctoral School in Philosophy and Human Sciences of the University of Milan (Italy). His research is at the interface between quantitative studies of science, philosophy of science, and science policy.

Gregory Rupik is a PhD candidate at the University of Toronto's Institute for the History and Philosophy of Science and Technology. He specializes in the history and philosophy of Romantic-era biology, and his dissertation tracks the influences of Romantic biology on contemporary shifts in evolution theory. Gregory has also worked closely with Hakob Barseghyan to establish and grow the community of scientonomy at the University of Toronto and is an editor for the peer-reviewed journal *Scientonomy*.

Caterina Schürch studied History of Science, Philosophy of Science, and Biology in Berne and Munich. She currently works at Ludwig-Maximilians-University's Department for History of Science on a dissertation concerned with cross-disciplinary research collaborations in the 1920s and 1930s.

Laura M. Sellers has a BA and MA in history and philosophy of science from the University of Leeds. She completed her PhD in 2018, also from the University of Leeds, which explored the impact of medicine and psychiatry on Victorian British convict prisons and sciences related to crime. Her research focuses on history of medicine, psychiatry, science, institutions, and crime. She currently works in museums and heritage.

Massimiliano Simons is an FWO PhD fellow at the Institute of Philosophy (HIW), KU Leuven. His main interests are situated in the sociology, history and philosophy of science, especially focusing upon recent shifts within contemporary life sciences such as metagenomics and synthetic biology. He has published on the philosophy of science and technology of Thomas Kuhn, Gaston Bachelard, Michel Serres and Bruno Latour.

Jinwoong Song is a Professor at the Department of Physics Education, Seoul National University, Korea. He took his BA in physics and MEd in science education at Seoul National University and received PhD in science education at King's College London (1990). His research interests cover a wide range of topics in science education, including the contextual dimension of physics learning and teaching, informal science education, socio-scientific issues, science culture, history of science education, and linking history and philosophy of science to science education.

Matteo Vagelli holds a BA and an MA in philosophy from the University of Pisa. He obtained his PhD in philosophy from the Université Paris 1 Panthéon-Sorbonne and the Scuola Alti Studi Fondazione San Carlo. He has done research at the University of Chicago, the Goethe Universität Frankfurt am Main and the Centre Marc Bloch in Berlin. He coordinates the research network 'Epistémologie historique: history and methods of historical epistemology'. In 2017 he held the Chair 'Pensées françaises contemporaines' at the European University Viadrina (Frankfurt an der Oder) and he is currently a post-doc at the FMSH (Paris).

Mark Thomas Young is a PhD candidate in the philosophy department at the University of Bergen and is currently working on a dissertation in the History and Philosophy of Science. His research concerns scientific instruments, craft practices and tacit knowledge in the history of science, with a particular focus on the early modern period.

Acknowledgements

We would like to thank the British Society for the History of Science, the British Society for Philosophy of Science, and the Centre for History and Philosophy of Science at the University of Leeds whose generous support allowed us to hold the conference which originated this volume. With their help, we were able to allow the early career speakers to register for free and we were also able to offer travel assistance to the speakers based outside of the UK.

We would like to thank Greg Radick and Jamie Stark for their feedback on the volume, and the academic community at the Centre for History and Philosophy of Science for the many conversations that have helped along the way.

We would also like to thank everyone who attended the forum and whose thoughtful questions and comments have helped shaped this volume. In particular, we would like to thank our keynote speakers, Jon Hodge and Chiara Ambrosio, for their kindness in attending the conference and for providing inspiring reflections on the past, present, and future of iHPS.

Introduction

In 2016–2017 the Centre for History and Philosophy of Science at the University of Leeds celebrated its 60th anniversary. Since 1956, the Centre has been one of the few institutions promoting the study of science from a combined historical and philosophical perspective (other such departments include the HPS departments in Cambridge, Indiana, and Pittsburgh, and the IHPST in Paris and Toronto, to name a few).[1] The editors of this volume viewed this anniversary as an opportunity to reflect on the legacy, the current state, and the future of Integrated History and Philosophy of Science (iHPS).

We decided to organise a conference with two main goals in mind. First, we wished to bring together researchers interested in the rich heritage of iHPS, the current issues that it faces and the new potentials for research that lie in wait. In the call for papers, we intentionally construed iHPS broadly, as the study of science from a combined historical and philosophical perspective, in order to be inclusive of a wide range of approaches and traditions (including those which do not label themselves iHPS, such as the historical epistemology tradition). Second, we wanted to offer better visibility to those who, like ourselves, were nearer the beginning of their career. We therefore explicitly aimed to showcase the works of early career iHPS scholars.

We received an overwhelming response to our initial call for papers. The high quality of the abstracts for presentations enabled us to host in January 2017 'The Past, Present and Future of iHPS: An International Postgraduate Forum'. Over the course of two days, early career researchers from across the world discussed theoretical ideas surrounding the relationship between the History of Science (HS) and the Philosophy of Science (PS), the fine texture of the case studies they were working on, the methodological specifics of their research, and a number of other topics. The keynotes, which were delivered by Chiara Ambrosio and Jon Hodge, took stock of the past and present conditions of iHPS, and were inspiring to the early career delegates interested in realising its future.

We have organised the volume into two main parts. The first part, 'Problematising the Relationship Between History of Science and Philosophy of Science', contains chapters emphasising and attempting to solve methodological problems to do with the relationship and the integration of history and philosophy of science. The second part, 'iHPS in Practice', showcases chapters which put the integrated approach into practice through various examples and case-studies.

Reflections about existing obstacles to the effective integration of HS and PS run through the first part of the volume. Several contributors note that there is a tension between the apparently incompatible approaches employed by HS and PS to study science, with the former supposedly taking on a descriptive approach to scientific thought, and the latter providing normative models for how science should develop. The chapters in this section attempt to resolve such tensions by proposing novel ways of thinking about the relationship between HS and PS.

In the first chapter of the volume, Greg Rupik outlines how scientonomy, an empirical science of science, provides an opportunity for a new method of integrating HS and PS. This centres on a general theory of scientific change first formulated by Hakob Barseghyan in his 2015 *The Laws of Scientific Change*. This approach seeks to combine PS and HS by proposing four laws that govern scientific change, and as such creating a metascience which can integrate HS and PS by bringing together the descriptive and normative approaches to the study of scientific development. Rupik argues then that HS and PS each provide the observational and theoretical components to this empirical meta-study of the study of science.

Caterina Schürch picks up on some of the problems the integration of HS and PS presents by asking whether iHPS is feasible and desirable, and considering both the arguments of those who see the project of iHPS fail, and the arguments of those who instead believe that history can provide insights into philosophical questions, and that philosophy can bolster historiographical analyses. Schürch argues that iHPS is the method of choice whenever we deal with metascientific problems that refer to interrelated philosophical and historiographical questions. A prime example of a research problem suited to iHPS is the issue of understanding past research practice. Schürch illustrates this point by confronting the historiographical work on the integration of physico-chemical and biological methods in the early decades of the twentieth century with some key assumptions of the new mechanical philosophy and argues that an integration of the two accounts would strengthen them both.

Claudia Cristalli addresses narrative-based explanations in the sciences, focusing on palaeontology and geology, the so-called historical sciences. She looks at how an integrated approach to narrative explanations is necessary to understand how this form of explanation functions. As in Rupik's chapter, Cristalli attempts to understand the ways in which scientific change occurs, and achieves this through an interrogation of the use of narrative to formulate and advance the research questions of these scientific disciplines. By carrying out a philosophical and historical analysis of the role of explanation in the historical sciences, Cristalli, with reference to the work of Hempel, Whewell and Pierce, claims that an integration of HS and PS is necessary in order to fully understand the role of narrative in the historical sciences. Cristalli concludes that this integrated approach can provide the beginnings of a narrative interpretation of explanation that can be applied to fields other than the historical sciences, and can be used more generally by iHPS scholars.

Eugenio Petrovich uses an investigation of science policy to address what Richard Giere called the marriage between HS and PS, which is parsed out by asking whether it was a marriage of convenience or an intimate relationship. According to Petrovich, this question is as important as it was when Giere first posed the question in the 1960s, and this translates itself into the identity, goals, and theoretical basis of iHPS. The chapter frames Kuhnian, neo-positivist and Popperian approaches to HPS in a Hegelian dialectic that respectively represent a thesis, an antithesis and a synthesis, and it is the logic of Popper's approach that is present in the formulation of science policy. Petrovich provides a series of case examples which demonstrate this, and goes on to offer a research methodology which can provide a new synthesis of HS and PS.

Picking up on addressing Giere's metaphor of marriage to describe the relationship between HS and PS, Matteo Vagelli in Chapter 5 claims that this is a problem for Anglophone HPS, and that French historical epistemology can provide important solutions that can help the integration of the two. The work of some of the principal figures in historical epistemology, namely Gaston Bachelard and Georges Canguilhem, predates that of the historical turn that was signalled by the publication of Kuhn's work. Despite some interest in historical epistemology from the Anglophone world, it has often been understood as being a form of Science and Technology Studies (STS), and the unique formulation of the research programme has not properly been considered. More particularly, Vagelli contrasts the naturalising trend prevalent in certain areas of the Anglophone debate with the 'normative turn' instantiated by Bachelard and Canguilhem. In highlighting the important differences between STS and historical epistemology, Vagelli brings light to a poorly understood movement in Anglophone iHPS, which, given proper treatment, provides insight into an effective integration of HS and PS.

In the third of our trio of historical epistemologists, Massimiliano Simons builds upon the outline of historical epistemology provided by Matteo Vagelli in order to take up Imre Lakatos's famous play upon Kant's phrase in the *Critique of Pure Reason*: that HS is blind without PS, and PS is empty without HS. Simons reaffirms Lakatos's belief that empirical HS problems cannot be resolved without some recourse to PS, and that philosophical questions need some empirical data to provide a field for experiment. Focusing on the work of Bachelard, Simons argues that French historical epistemology can help us develop a finer grained understanding of what the relationship between HS and PS could and ought to be. Bachelard's work is important because it provides a methodological framework which prioritises the practice of science over the philosophy of science in attempts to understand how scientific research operates but at the same time is not afraid to make normative judgments about the ways in which scientific research ought to operate. This tradition still exists in the work of current French epistemologists, and Simons show how this innovative method of integrating HS and PS is demonstrated in the work of Isabelle Stengers.

The chapters in the second section put the integrated approach into practice by presenting cases which, while often well-known in HS or PS scholarship, strongly benefit from being reevaluated from an iHPS perspective.

In the first chapter in the section, and the seventh in the volume, Mark Thomas Young draws upon the existential phenomenology of Martin Heidegger in order to investigate the relations between craft knowledge and the emergence of experimental research during the early modern period. Young calls into question the strong continuities that have been drawn between the two, and utilises Heidegger's ideas in order to understand how the development of decontextualised knowledge guided the development of experimental practice and theory by the Royal Society, and although feted as being a break with craft-based forms of knowledge, in fact played into an epistemological tradition that had existed within Western philosophy since antiquity. Young's chapter then seeks to provide a picture of iHPS that uses existential philosophy to provide a valuable insight into a historiographical debate from HPS, and thereby provides a powerful model for iHPS.

Andrea Gambarotto draws upon recent work carried out by Uljana Feest and Friedrich Steinle to provide a conceptual history of teleology. Feest and Steinle argue that iHPS has two tasks: first, it outlines the historical processes by which concepts of scientific thought are developed in order to demonstrate the evolution of scientific knowledge, and second, that it creates philosophies of science which provide a faithful treatment of the ways in which scientific practice is actually conducted. This is in contrast to the normative models of science that are often proposed by philosophers of science. Taking this lead, Gambarotto proposes a history of the concept of teleology that commences with the early modern conception centred around Cartesian mechanism, before harking back to Aristotelian thought in order to uncover an archive of theoretical alternatives to those that we have become accustomed to in the modern period. The final section of the chapter then provides a case study that outlines how Hegel drew upon the Aristotelian conception of teleology in order to demonstrate how the history of science can be employed in the service of the philosophy of science. In this way then, Gambarotto provides a model which enables conceptual histories and the history of ideas to provide fuller accounts of the concepts used by philosophers of science, providing modern examples of the related concept, purposiveness, from the work of the cyberneticists Humberto Maturana and Francisco Varela.

Picking up on this thread of the recent work in the history and philosophy of cybernetics, Joe Dewhurst provides a history of the cybernetic origins of enactivism and computationalism. Dewhurst seeks to question the preconceived notion that these two schools of thought are in opposition by carrying out a philosophical analysis informed by a historical charting of the origins of the two schools to the first order cybernetics developed by Norbert Weiner and a number of other individuals. Dewhurst claims that during this time, the notions of biological homeostasis and neural computation were able to co-exist, but goes on to state that the two schools diverged due to the development of the enactivist notion of autonomy, which although having its roots in biological homeostasis, found its development through Maturana's autopoetic theory to the modern enactivism. In tracing this history, Dewhurst seeks to argue that the enactivist notion of autonomy is incompatible with computationalism if it is understood as a semantic phenomenon, and that by looking at the history of both schools, it is possible to reconcile these opposing philosophical schools.

Klodian Coko examines the well-studied but often misrepresented case of French physicist Jean Perrin's argument for molecular reality through the lens of a Hermeneutic-Historicist approach to the integration of HS and PS. This approach provides Coko with the framework to move beyond the conflicting philosophical interpretations of Perrin's work and to analyse and contextualise important structural elements of Perrin's argument, uncovering that it was based on the employment of the epistemic strategy of multiple determination. In addition, Perrin's case can be used to develop a conceptual framework for dealing with the structure and epistemic import of the multiple determination strategy in general. On the one hand, this conceptual framework can be used to understand the structure and epistemic import of other cases of multiple determination from past or current science. On the other hand, it can be enriched and further developed in contact with historical material. Coko therefore argues that a Historicist-Hermeneutic approach paves the way for a 'mutually beneficial' interaction between HS and PS.

The final two chapters tackle the issue of pluralism, in historiography and science education, respectively. Alex Aylward argues that when we possess several differing historical accounts of the same scientific episode, they are often viewed as 'competing'. The persistence of historiographic pluralism with respect to any particular case-study is usually conceived as an obstacle to be overcome in pursuit of the (one) 'true' historical account. Using a case study from the London Royal College of Surgeons he urges that we adjust our attitudes to pluralism in the History of Science, in response to lessons from the perspectivism movement in the Philosophy of Science. We should actively pluralise our historiographical perspectives upon particular scientific episodes, in the pursuit of greater completeness, along with a host of other historical and philosophical benefits.

Wonyong Park and Jinwoong Song's chapter concerns the relationship between science education and iHPS. Sparked by decades of scholarship in science studies, 'science as practice' has recently begun to attract growing attention from science educators, who find teaching 'the scientific method' to no longer be valid in school setting. Recent curricular reforms such as the US Next Generation Science Standards also support such a 'practice turn' by proclaiming the teaching of scientific practices as their key objective. Consequently, philosophers have come to notice that there rarely exists a single correct account which fully explains the entire natural phenomena, but more commonly found is a plurality of theories, models, and explanations that are often incompatible with one another. The authors therefore set out to make a case for using iHPS to rethink educational practices, by examining how realist forms of scientific pluralism illuminate the dilemma between realism and constructivism in science education.

With these twelve chapters, originally papers delivered by early career researchers at our conference, this volume intends to contribute to the further advancement of iHPS by providing a snapshot of some of the most recent developments in iHPS scholarship and gesturing optimistically toward its future.

Note

1   Although iHPS has been relatively recently established as an institutionalized field of research, at Leeds and beyond, one could trace the origins of the integration of historical and philosophical considerations about the study of nature as far back as Aristotle. This prehistory of iHPS is explored in the preface to this volume, which is comprised of an interview with the first of our keynote speakers, Dr. Jon Hodge, who is a long-time fixture within the Centre for the History and Philosophy of Science at the University of Leeds.

Origins, trends, methodologies and divisions – reflections on the past, present and future of iHPS: A keynote interview with Jon Hodge

Before the iHPS forum the editors sat down to chat with Emeritus Fellow Dr Jon Hodge who has been based in the Division, later Centre, for HPS since 1974. Jon still offers insights to many students passing through Leeds and offered his thoughts on the past, present and future of iHPS for the forum. This interview was presented as a keynote at the forum and what follows below are revised excerpts from the conversation he had with the editors of this volume.

0.1 The origins of IHPS

0.1.1 Prefatory warnings

Four comments in advance: first, I am often drawing on unreliable memories here, memories sometimes tracing to rumours and gossip; second, while I have some credentials as a professional historian of science, my philosophical and social studies credentials are amateurish; third, I have long been aware that the relation between history and philosophy of science has been a disputed topic discussed by such people as Larry Laudan and Ron Giere: but I have only very recently learned about the current issues associated with the labels 'hyphenated HPS' and 'integrated HPS', and addressed in publications by Hasok Chang and others and taken up at our conference. Finally, let me take this chance to thank the conference team for giving me this interview opportunity and for valued help in revising the original transcript.

0.1.2 What were the origins of the field, when and why did it come about as a field, and could you then discuss some of the reasons why this occurred specifically in Leeds?

Those are challenging questions. You used the word 'field', and I think that that is appropriately vague. If we ask when did hyphenated HPS become a profession, then we'd be talking really about the last fifty or sixty years; that is when there were first standard ways to get trained in hyphenated HPS, that is when there have been programmes officially devoted to its studies and hirings in it; and you might say that another word comes into play here, the word 'discipline'; for yes, we've had a discipline of hyphenated HPS for the last fifty, sixty, seventy years; but in a broad sense, as a topic rather than a field or a discipline, hyphenated HPS goes a long way back; you can make a good case for Aristotle doing hyphenated HPS, for when he gets into a number of questions about science, he asks what are the opinions of the many and the wise, and how long have various beliefs been held and by whom and for what reasons and so how much credence to give them and so on.

When people became self-conscious about modernity, around the time of Isaac Newton, a famous controversy broke out between the ancients and moderns, that ushered in another way of integrating the history and philosophy of science; because people had theories about how progress takes place in the sciences, and they wanted to say that there had been progress in the modern age, progress beyond where the ancients left things.

Then you fast forward again to the 1830s and you have Auguste Comte in France and William Whewell in this country [the UK], who were really developing general theories of change and progress and reinforcing those by drawing upon philosophical resources. Very strikingly, Comte seems to draw upon English and Scottish resources, John Locke, Francis Bacon and David Hume; while Whewell, although he certainly draws upon Bacon, draws more than anyone else upon Immanuel Kant.

In so far as Whewell has been a father for HPS in the English-speaking world, he's been a German father; whereas Comte, the father of a lot of *épistémologie de la vie* in France, was more English and Scottish. So, the nationalistic issues surrounding the origins and boundaries of iHPS (integrated or hyphenated HPS) are complicated.

So, it's probably generally agreed that hyphenated HPS was not recognised as an academic specialty, discipline and professional category, in the English-speaking world at least, until the 1950s. And it was mostly understood as drawing on Germanic philosophical inspiration, especially Hegel and Kant (and the later Wittgenstein) and in its Hegelian and Kantian alignment it was seen to be in opposition to the dominant analytic philosophy of science, logical positivism.

0.1.3 What were the origins of iHPS in Leeds?

It helps here to focus on four people: Mary Hesse, a Protestant Christian mathematician – she was in our maths department in the 1950s; Ted Caldin – a Roman Catholic chemist; and Stephen Toulmin, a boy wonder, who had studied at the feet of Ludwig Wittgenstein during Wittgenstein's later years. (Wittgenstein, rumour has it, only had a chair for himself in his room; his students had to sit on the floor at his feet.) At Leeds too, there was Peter Alexander who was a historian of philosophy, particularly of Locke, and who was fascinated by Locke's debts to Robert Boyle and the new mechanical philosophy of the seventeenth century. Lately I have been reading a marvellous book on the history of atomism from Democritus to Newton by his protege Andrew Pyle

Together, these four people lobbied to get HPS going, and it was given a home right here in this department in Leeds presumably because Toulmin was head of the department at that time …

0.1.4 On first arriving at Leeds HPS…

Geoffrey Cantor and I came just after the 'Golden Age', and the Golden Age people were really a hard collective act to follow: especially Charles Webster, Ted Maguire, Maurice Crosland, Charles Schmitt, Piyo Rattansi; they were all at Leeds in the years before 1974 – which is about when Geoffrey and I arrived. And they had gone off to various prominent positions around the world; but, notoriously, there was one not so good thing about those golden years: there had been factions, I'm sorry to report; there was a polarisation and the gossip was that if the Leeds group were in the pub, then there were two tables (I won't name names), such were the divisions.

But peace had broken out when Geoffrey and I arrived, not because we were peacemakers, but because some of the more divisive folk had gone. There was a real ideological issue in their divisions. One cluster was very much for historical scholarship, and the other cluster wanted to be politically engaged, and take up green issues and issues about freedom and oppression in scientific life and so on. But, as I say, that division did lessen, and peace is still with us, I am glad to report. The other point I would make is that the operation was in the mid-seventies very small; there were only very few students and a handful of postgraduates at any one time. As for teachers there were Jerry Ravetz, Geoffrey Cantor, Bob Olby, John Christie and me: just a team of five, whereas the number today would be twelve or fifteen – there has been a huge increase. Then when Jerry retired, we were down to four people and were so when we collaborated in producing what we called the 'Leeds Companion' to the history of modern science. So, this expansion, in the last twenty years, is hugely welcomed by people like myself who can remember those pinched and struggling years.

0.1.5 Can you tell us about the balance between History of Science and Philosophy of Science in Leeds HPS?

The programme at Leeds got off to a rather lop-sided start. It's probably true to say that Steven French was the first fully qualified philosopher of science to teach HPS at Leeds: prior to that, philosophy of science was taught to undergraduates, but it was taught by historians like myself who were amateurs and self-taught and were not doing research in philosophy of science. And it is good to see that the balance is better now, even Stevens – sorry for the joke; and it's been made even more even, now that Ellen Clarke has arrived, a specialist philosopher of biology whose first encounter with her special field was probably as an undergraduate at Leeds in an amateurish course of mine.

0.2 Trends in iHPS

0.2.1 Do you think there are trends in iHPS?

Trend is a good word, because it has a serious meaning, and we know that there are trends. But it also has a slightly derogatory meaning, where to say that someone is being trendy is a bit of a put down. I was once told that trends in HPS last around eight years, but of course, some, not always the best ones, last quite a lot longer than that.

I'll give you an example of a trend: twenty or thirty years ago, people started talking about the body, and there was a volume of essays put out on science incarnate, a volume all to do about how the bodies of scientists influenced their practices, and obvious examples. Dalton was red-green colour blind, and maybe this makes a difference to the way that he did science. I think that body-language trend burned itself out in around eight years, and now it sounds rather old fashioned to talk about the body. The language came much more from history of science than philosophy of science: people talked a lot about the *body politic* in the 1970s and body talk became fashionable in a number of areas, and I think that this is a fashion or a trend that has gone; it was useful in its day.

I would say the biggest long-run trend that I have witnessed, is the decline and fall of positivism in the English-speaking world. Of course, the foremost form of positivism that was dominant in the English-speaking world was logical positivism, Vienna circle positivism, that really was very dominant in the 1950s: it dominated in the philosophy of science and was influential in the history of science. Logical positivism was analytic philosophy of science and had arisen partly as part of the Gottlob Frege-inspired analytic swing away from late nineteenth century Hegel-dominated idealist philosophy.

A leading logical positivist was Rudolph Carnap, one-time student of Frege. He once spent some time at Harvard; and, legend has it, when he arrived, Bernard Cohen did the decent thing as the main man there in the history of science and invited Carnap to give a talk in one of their History of Science seminars; and Carnap, who was one of the nicest guys ever, said spontaneously that he would be very happy to do that. Also, one of the most honest guys ever, a few days later he got in touch with Cohen and said that he should not have accepted the invitation because he had no interest in the history of science. And that was probably around the mid-50s. Now, fast forward ten or fifteen years only, and almost no young philosopher of science would say that he or she had no interest in the history of science; and it is well known which Anglophone people were responsible for that shift: Stephen Toulmin, Thomas Kuhn, Russell Hanson, Paul Feyerabend, Imre Lakatos, and several others who intrigued and provoked logical positivist philosophers, and gave them something new to think about. Most of these historicist HPS pioneers had of course done important historical case studies and had theories about the long run of progress and change in science, and it quite quickly became widely thought that it was a weakness of logical positivism as a philosophy of science, and of analytic philosophy itself, that neither had much to say about progress and change, and those historical issues.

Even now, I find myself waging war against, if not logical positivist, but definitely broadly positivist views about Darwin, for example; and it was well said by Hilary Putnam that positivism is science's philosophy in that it is the philosophy that scientists love best. And that's no coincidence: it was designed in the nineteenth century to legitimate the new profession of science, and it does it in a very self-congratulatory way by holding that science has more authority and more scope than anything else. To put it crudely, as a positivist you could really claim that there is science and there is rubbish, and you are either doing one or the other. Only scientists should be judging and planning science, and that's music to scientists' ears and often leads to a very triumphalist, internalist and Whiggish history of science. I won't name names, but I would be prepared to say that there are a number of people of good reputation who are Darwin buffs and who are still listening if not dancing to that tune.

But of course, among professional philosophers of science, positivism in all its forms, including logical positivism, has simply not been a career option for decades now. There was reputedly a famous moment I think in the early seventies, when someone, Clark Glymour I believe, was accused at a philosophy of science meeting in the USA of being a logical positivist, and he stood up and announced that he was happy to be labelled a logical positivist, and the whole room rose and applauded, not because they thought it was a good thing, but because it was a gutsy move to make at the time.

So, yes, I would say that this movement away from logical positivism has been a very big trend and consequential change, as is evident from the attention now given by historians of the philosophy of science to the rise and fall of that whole way of thought.

0.2.2 Do you think there are geographic and linguistic influences upon trends within iHPS?

There's a way of looking at this question which is geographic and historical. I am prepared to say that after about 1800, all Western philosophy has been predominantly Germanic, and that includes Austrian. And so, what are the great divides? Well we are often told that there is a great divide between continental philosophy and analytic philosophy. In fact, insofar as that is a division, it really is one that exists within Germanic traditions. To put it in a nutshell, what we call continental philosophy looks more to Hegel and Friedrich Nietzsche than it does to Gottlob Frege and to Moritz Schlick. What we call analytic philosophy looks to this latter pair of figures. All those are Germanic names, and I'm sorry to say things which dent Anglophone self-respect and indeed French self-respect, but if you look at the big names of French philosophy, then they are all drawing massively on Hegel, Nietzsche, Husserl and Heidegger and so on, and throughout the last century in Paris they have almost all agreed that they don't want to know about Frege and Schlick and Carnap, whereas in the Anglo world, Frege and Wittgenstein, another Germanic name, they are the fathers; and I would say that a big shift in the English speaking work is that it no longer costs you career points to have a copy of Nietzsche sticking out of your briefcase, as it would forty years ago, when I first started working in this department.

0.3 Different traditions and methods within iHPS

0.3.1 What are your thoughts on iHPS's traditions?

I've been impressed when looking at some of the recent literature on integrated HPS, that there is what I think is a very healthy pluralism. Let's take a book like Jean Gayon's book, *Darwin's Struggle for Survival*: Jean is the grandchild of Canguilhem, because intellectually he's a child of François Dagognet who was Georges Canguilhem's pupil. But I'm reliably informed that Imre Lakatos was a considerable influence on that book. Who was Lakatos? Well he was Popper's successor, and you could say that he was a Hungarian Popperian, but he was also a Hegelian due to his Hungarian education. And as a Hegelian he was very much a historicist; he was famous for saying that all theories are born refuted, and they need to get over those initial refutations, and that's how Jean tells the story of the theory of natural selection – that it was not a reasonable theory to accept in 1859, it only really becomes reasonable to accept it in the twentieth century: so then why did it survive until it could become acceptable, and why didn't it die in its first refuted form? The answer is that people thought it had promise, and they were working on refuting the refutations, but they only succeeded on refuting the refutations in the twentieth century.

Now that is an example of a historiographical-philosophical tradition, if you like, and a stance that is rather different from anything Whewell ever came up with, and anything that may have been done in the English-speaking world. Jean Gayon's stance is an interesting synthesis, if you like, of Lakatos's combination of Germanic Popperianism with a residual Hungarian-Germanic Hegelianism and the French tradition influenced by Canguilhem.

0.3.2 What do you think about methodologies within iHPS?

I've struggled with this issue of methodology; I've not signed up with Feyerabend in being against method, but it's not a word I find myself applying to my own work, or the work of other historians, except in a very general way. I suppose two people could be said to differ in their method in integrated HPS, if one of them doesn't really talk about people, doesn't really talk about authors, doesn't really talk about actors, but talks about concepts and texts, whereas the other person does talk about authors and actors and so on. That's in a broad sense a methodological difference, a matter of what you are going to let in to your account. And I'm on record as saying that I find the Canguilhemist tradition impoverished because it does not really allow for people, and institutions, and ideologies, to come into the story: in that tradition, the history of science is the history of concepts, that is the mantra. And if that's a method stance, then I must say I find it that's an impoverishing one.

More and more I've come to the feeling that if methods carry with them prohibitions, then I'm against those methods; it seems to me that one thing we learn from doing the history of science is that sooner or later we may need all sorts of things to come in. In understanding why high energy physics went the way it did in the 1950s in the US, then you have to take the Cold War into account, and if we are interested in why Darwin went on the *Beagle* voyage, we have to look at who paid for that voyage (ultimately the British national government) and what they were going to get out of it (informal imperialist and hegemonic commercial advantages). These are old fashioned, even vulgar Marxist questions, and none the worse for that; for if we just say that a Darwinian text came out of that voyage, then that's very impoverishing. Camille Limoges, when a young Canguilhemist, wrote a remarkable book called *Natural Selection, The Constitution of a Concept* in which he goes through Darwin's notebooks as though they weren't written by anyone in particular, and they are just there and you can just watch this concept being *constituted* towards the end of notebook D, and it has nothing to do with who Darwin was inspired by or talking with at the time, what his ambitions were and who he was trying to impress and who he was trying to discredit. In other words, it's history of science with the people left out. I don't like leaving the people out. So, I have disagreements with historians who are wary of including stories about influences and intentions in their histories of science, because I think influences and intentions and especially influences on intentions are what make people interesting and intelligible. It's like good journalism which tells you on Sunday why some politicians did something on Tuesday. They were influenced in certain ways and had certain ambitions and goals, and you realise this is beginning to make sense. To me, if we are trying to make sense not of what some politicians did a few days ago, but what some scientists did several hundred years ago, then again, we want to take the influences and intentions into account. Post-modernism and poststructuralism have taboos across those areas, just as analytic philosophy and behaviourism and eliminative materialism do.

0.3.3 What are the relationships between methods within iHPS and other branches of history?

We always have to remember that many historians of science work in history departments. I've worked in history departments, and there's quite a lot of pressure to recognise some of the traditions. For instance, in history departments that I was in in the 1960s and '70s, the *Annales* view of history was hugely influential, and Ferdinand Braudel was the master. And it was inspiring stuff: one took the long run into account, and I found myself using that phrase, because I had been asking how did Western philosophy and science get from Plato to Darwin – that's a long run question. It's a question that philosophers might come up with, but this business of the long run is associated for historians with the *Annales* school of history rather than with any philosophical school.

There was of course a turn away from the long run of history during the 1970s and '80s. I remember once having lunch with a postgrad here and he was doing a study of nineteenth century Northern English natural history, and I reflected that he didn't seem to be interested in what social classes these people belonged to, and which social classes were losing, and which were winning power at this time. And he told me that his supervisor told him that he shouldn't be interested in those questions because those questions were looking unprofessional in the 1980s. They looked like the kind of things that were written about by retired journalists who wrote popular history, big picture stuff that had class interests included. I'd like to think that this professional stance is no longer influential, and that to look professional you don't need to be very narrowly focused and leaving out grand narratives and big pictures.

0.4 Are there polarisations and divisions within iHPS?

You could say that there is a contrast to be drawn within the history of science: some historians, when they've done their work, and it may not be deliberate, depict the authority of science as diminished, because science is shown to be subject to political or religious influences. To use the inevitable word, when these people have done the history of science, relativism is the bottom line. Of course, by contrast, we have people who are openly philosophically motivated, and wanting to say: 'No, you can immerse yourself in the history of science as a philosopher and still come out a realist and a confident progressive realist', insisting that we can really show that we really do know more about the world out there than they did in King Charles's or Queen Victoria's time.

There's no question that that's a polarisation, but I don't see that being acted out in the politicised way that went before. I remember a conversation in the early 1970s with a visitor to Pittsburgh, and we asked him why he was such an externalist – of course that was the buzzword then – and he said that we in the West are in a terrible perverse war in Vietnam, and he can't go there and demonstrate on the streets, but at least he is trying to discredit this deference to science which is a component of the ideological and military warfare that is being waged in Asia. It was a salutary moment: he was trying to make us internalists (and I was one at that time) look reactionary and in denial about big issues of liberty and humanity and so on across the world. I have never quite bought the argument, but I feel it is one worth bearing in mind, even if HPS people don't often get into these kinds of discussions these days.

There was a version of that kind of thing in Paris in the 1970s, where if you took the intentions of an author seriously, then you were implicitly an authoritarian, and were probably far too tolerant of people like Francisco Franco, and far too tolerant of racism and so on – and you think: 'hold on a minute, I'm taking seriously Darwin's authorial intentions in writing the *Origin*. Does that make me politically out of order?', and the answer at the time was yes, potentially it does. I think then that that kind of politicisation is no longer prevalent, but then I may be wrong: perhaps I no longer have beer with the right people in the evenings at conferences.

0.5 How do you see the future of iHPS?

Looking to the future: I'm optimistic because some views which were unhelpful no longer prevail. Some historians of science, in the late 1960s and '70s, said there really is a choice: we can assimilate ourselves to social theory, or we can assimilate ourselves to philosophical theory. Philosophical theory, they said, is outmoded, and probably illiberal: social theory is active and less oppressive, and we should go that way. The so-called strong programme of sociology of knowledge in Edinburgh took that view. And young people at that time took the view that they needed to decide between deferring to the philosophers or to the social theorists. At that time, I was at the University of Pittsburgh, and HPS there was very much signed up to Integrated HPS. At the other end of the state in Philadelphia there was a programme headed by the expat Yorkshireman, Arnold Thackray, which was very much signed up to the sociological stance, and I think Steve Shapin was there at the time.

There was polarization, which we are mostly far away from now. I think Steven French will confirm this: that philosophers of science will not lose their licence today if they are found to be saying something friendly in a footnote about sociology of knowledge. Equally, social historians of science, or social theorists of science, don't lose their friends and their licences if they make some points about the reductionist tradition in the philosophy of science, and how it influenced molecular biology in the 1970s. And that is I think very healthy, and it is one reason I am very optimistic about the future. It could be of course, that things take one of those unhealthy turns in the future, and young people are told that they had better decide which side they are on, this side or that side, and if it's this side, our side, then here's the list of things you don't do.

To speak crudely once more: some history of science leaves out the people, some leaves out the science. Neither of these omissions looks good and for two obvious reasons. First: much of the interest many of us have in the history of science is there because we are intrigued by the challenge of understanding both the people and their science by relating them to each other; and second, in attempting this we are fascinated by the extraordinary mix of things that may bear on that challenge: institutions, class structures, experimental instrumentation, formal training, informal mentoring, family backgrounds, ideological and metaphysical alignments, forms of economic life, fieldwork, lab work, mathematical skills, personality conflicts, war and peace, races, empires, etc., etc.: the inventory is endlessly extendible. And, because it is, the opportunities for hyphenating integrations of HS with other academic fields and disciplines are numerous and diverse and all the better for that.

As aphoristic contributions to future prospects of healthy pluralistic stance and practices, let me offer, with apologies for their banality, a couple of further reflections. First, those people are right who say that just as contexts can enlighten texts, so the clarification of texts can point us to pertinent contexts. So textual and contextual concerns are mutually enhancing and not competing alignments. Second, let lots of hyphenations bloom. As implied already, it may be appropriate to move beyond the singly hyphenated HPS to multihyphenated ambitions, programmes, stances and so on.

As an example, already crying out for multihyphenated recognition and labelling, let me mention one of my favourite societies. Over the last three decades or so I have gone to about half the biennial meetings of the International Society for the History and Philosophy and Social Studies of Biology. It is a very comprehensive, pluralistic, organisation, and all sorts of people come with all sorts of interests, with all sorts of methods, points of view and axes to grind but all concentrating on biological topics and issues. At a minimum the label for what is promoted by ISSH, as it is known, would have two hyphens as in History -Philosophy-Social Studies of Biology.

I'd like to see even more multi-hyphenating. And I am sure I will: sure, thanks, for example, to reading recently John Zamitto's multihyphenated book – *A Nice Derangement of Epistemes* – surveying critically historiographical, philosophical and social theoretical views of science from Quine to Latour. Among other favours this book has done me it has confirmed my feeling that being counter-suggestive can be useful sometimes. When some trend in History of Science has been fashionable I have usually reacted negatively; but then later when it has become unfashionable I tend to view it more positively, and to value it for emphasising important issues that have since dropped out of sight as more recent fashions have come to dominate. I have for example been through this cycle regarding the strong programme in the sociology of knowledge which I now, unfashionably, find useful in countering what I regard as some over-rated current fashions.

I have misgivings about academics whose identity depends too much on what they exclude, what they don't do, because I think that's unhealthy and unhelpful, especially in relation to teaching and to our public presentations of ourselves. It sounds banal, but I agree with those who hope things can become more, not less, inclusive. And I think that there are plenty of people around with that attitude, and that this will keep integrative, inclusive iHPS going long after its original Cold War motivations or its (Charles Percy Snow) bridging-the-two-cultures motivations have passed. It's here to stay: I don't fear for the future of iHPS. I reflect for example on Leeds HPS and on ISSH and on their strong links with each other, and, in reflecting on that and on much else besides, I am very optimistic.

Part 1

Problematising the relationship between history of science and philosophy of science

1 Scientonomy: A bold new vision for an integrated history and philosophy of science

Gregory Rupik

In 1989 Larry Laudan penned a retrospective for the journal *Studies in the History and Philosophy of Science* appraising the state of the field of HPS twenty years after he and Gerd Buchdahl had together founded *Studies*. Laudan concluded that, while philosophers had generally come around to granting that a historically-informed philosophy of science is a valuable enterprise,

> many (perhaps most) professional historians of science have refused to see the point. Indeed, the distance between mainstream history of science and the philosophy of science is probably greater now than it has ever been, notwithstanding it being the case that many historians of science still take philosophical issues seriously.
>
> (Laudan 1989b, p. 12)

The chasm that Laudan described separating the history of science (HS) from the philosophy of science (PS) arguably remains an institutional, disciplinary, and methodological reality today. That is not to say that this unintegration has sat comfortably with everyone. Since Laudan wrote his remarks in 1989, a number of impressive studies have defiantly bridged the HS/PS divide, in addition to special issues, edited volumes, conferences, and communities dedicated to a 'hyphenated' or 'integrated' history-and-philosophy-of-science (iHPS).[1] Despite these efforts, however, no clear consensus has emerged either about (1) what an ideal iHPS should look like or, more importantly, (2) how an iHPS might address the historical reasons for its unintegration.

This chapter seeks to address both of these issues. In what follows, I propose that an empirical *science of science* can provide a fresh approach to the field of HPS, capable of fruitfully integrating key components of both HS and PS.[2] As evidence for this proposal, I will introduce the work currently being done by a community of scholars who have taken Hakob Barseghyan's (2015) *The Laws of Scientific Change* as a starting point and the theoretical basis for an empirical science of science named *scientonomy*. My goal is therefore not only to articulate how scientonomy integrates HS and PS on a theoretical level, but to demonstrate precisely how this integration has already been operationalised in the scientonomy community. A fundamental strength of Barseghyan's work is that it offers a compelling historical hypothesis for why HS and PS are unintegrated, and crafts a theory of scientific change that explicitly redresses the causes of this unintegration. To make the case for scientonomy as an iHPS, I will begin by elaborating on the aforementioned historical hypothesis, arguing that the distance between HS and PS in the 1960s was due principally to, on the one hand, the conflation of normative methodologies and descriptive theories of scientific change in the philosophy of science, and, on the other hand, the gradual acceptance of the *dynamic method thesis*. Next, I will sketch the ways in which scientonomy theoretically integrates features of HS and PS. Finally, I will provide practical examples of how the community of scientonomy has crafted an iHPS.

1.1 HPS: integration and unintegration

By the 1960s, universities from Princeton (1961) to Toronto (1967) were enthusiastically establishing departments for HPS, gathering historians and philosophers of science together under a common administrative roof. Despite sharing office space and funding, the label 'HPS' did not entail a shared approach to the history and/or philosophy of science. Indeed, while some scholars like Karl Popper, Imre Lakatos, Stephen Toulmin, and Thomas Kuhn laboured to weave history and philosophy of science together in their own ways – at least *aiming* at an integrated history and philosophy of science (iHPS) – HPS as a field seems never to have necessitated this approach. Ronald Giere (1973), among others, posited that the 'marriage' between HS and PS was nothing more than a 'marriage of convenience'. Kenneth Caneva's recollection of Thomas Kuhn's HPS department in Princeton in 1967 could just as easily have been spoken by a graduate student in an HPS department today:

> There was almost no contact between the [HS and PS] parts, let alone fruitful interaction. And no one seemed to care. When I think back on the situation, I suspect a tacit but strong attachment to a preoccupation with fostering a proper professional identity may have played a key role.
> (Caneva 2012, p. 49)

Kuhn (1977) himself went further and argued that HS and PS cannot be practised at the same time. And while many today would doubtless agree that Imre Lakatos' (1978, p. 102) creative appropriation of Kant – 'Philosophy of science without history is empty; history of science without philosophy of science is blind' – remains an excellent motto for any iHPS, few if any would agree that Lakatos' strategy of rational reconstruction is a historically or philosophically adequate means of integration.[3]

In the following subsection I will argue that a principal cause of the typical unintegration of HS and PS has been the conflation of two distinct philosophical projects: the search for a *descriptive* theory of scientific change and the search for a *normative* methodology of science. To do so, I will first briefly describe the centrality of the *static method thesis* in the HPS practised from the end of the nineteenth century through to the 1960s. Second, I will demonstrate how the gradual acceptance of the *dynamic method thesis* posed a pivotal problem for the construction of a general theory of scientific change. Third, I will consider the implications of the dynamic method thesis and how the fate of both the normative and descriptive philosophical projects evolved. Finally, I will explain why the normative and descriptive projects have been uncoupled and will explore possible avenues forward.

1.1.1 The static method thesis: HPS before 1960

Generally speaking, historians and philosophers of science in the early twentieth century viewed science and its history as unique, progressive, and valuable. Some philosophers looked to science's history to help them understand precisely what features of scientific inquiry had allowed science to become such a successful knowledge-generating endeavour. Late nineteenth century philosopher William Whewell (1840, p. 1) characterised this early philosophy of science in *The Philosophy of Inductive Sciences*: 'The Philosophy of Science ... [is an] insight into the essence and conditions of all real knowledge, and an exposition of the best methods for the discovery of new truths.' For Whewell – and for many after him – PS's task is both descriptive *and* normative: it uncovers and clarifies the essence of the scientific method, and ultimately proposes it as the *best* way of evaluating theories and establishing new truths.

Pre-1960 HPS tended to understand the Scientific Revolution as the widespread employment of this *best* method of theory evaluation, *the* so-called scientific method, which they maintained was the cause and guarantor of scientific progress.[4] While the precise criteria of the scientific method were debated among those in the HPS community, most scholars agreed that it was some form of hypothetico-deductivism, requiring empirical confirmation of theories' predictions before those theories could be accepted. Notably, while it was believed that the scientific method had historically emerged and spread in the early eighteenth century, the method itself was understood as an epistemological means of justification with *universal* purchase, and this is demonstrated by science's progressive history. In other words, while they agreed that the hypothetico-deductive (HD) method of science may have debuted definitively during the Scientific Revolution, any genuine advance in knowledge through history was thought to have been due to its employment. This explains why traces of the HD method were sought even in the works of medieval and Early Modern scientists such as Avicenna or Galileo. As such, the scientific method was understood as static (fixed, transhistorical), and as central to gaining justified knowledge *today* as it had ever been. The adoption of this *static method thesis* had three major consequences:

1   Since the method was understood to have remained fixed (that is, *outside* the process of scientific change), explaining the changes in science amounted to explaining the changes in scientific *theories* alone. It was the criteria of this fixed method which were conceived of as the motor which drove scientists and their communities to accept some theories and reject others. The challenge of crafting an adequate theory of scientific change, therefore, amounted to discovering and accurately articulating that singular, universal method of science.

2   Historical episodes of scientific change could theoretically inform and correct the philosophical articulations of the method of science, and philosophers' formulations of the method could help shed light on the logic of certain transitions in science's history. Indeed, explaining any historical transition from one theory to another could be done by understanding how the criteria of the static method of science were employed by the scientists within the exigencies of a specific historical situation. (See Lakatos 1978.)

3   Philosophical attempts to articulate the fixed method of science produced a number of theories of scientific method – *methodologies*. These methodologies were meant to be both *descriptive and normative*, since they were supposed to both describe the criteria of the fixed method of science and prescribe the same criteria as the ones that we *ought* to employ if we want our knowledge to continue to advance (Nola and Sankey 2000, pp. 8–11).

1.1.2 The dynamic method thesis

By the mid-1960s, however, the foundations of the static method thesis had significantly eroded. The works of Kuhn, Ludwik Fleck, and N.R. Hanson, had begun to suggest that scientific change was not restricted to changes in theories, but rather that the expectations and criteria for theory assessment of science, that is the scientific method, *had also changed through time*. Later scholarship, from Laudan (1984), Paul Feyerabend (1975) and Dudley Shapere (1980), suggested that change in methods was almost as ubiquitous in the history of knowledge as change in theories.

Historical investigations thus gradually concluded that the scientific method, far from being a set of immutable criteria that drove scientific change, was itself dynamic and changeable, located *within* – and therefore subject to – the process of scientific change itself. The building case for *the dynamic method thesis* raised issues for many philosophers and historians of science. After all, if there is not one set of criteria which transhistorically constitutes the *logic* of scientific change, then what could guarantee that theories change according to any logic all? Faced with the *dynamic method thesis*, one had three major options (Figure 1.1).

While some accepted that there is no fixed and universal method of theory evaluation, many philosophers stuck to a version of the static method thesis (Option 1). Among many others, John Worrall insisted that despite all changes in specific methods (such as the transition from single-blind to double-blind trial

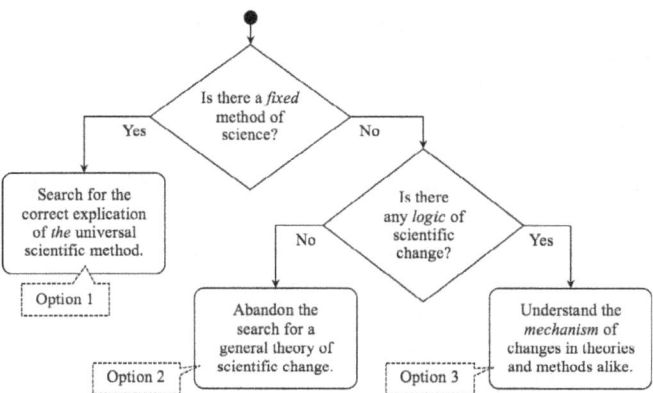

Figure 1.1 Dynamic method thesis

method of drug testing presented below) there exists a more fundamental method which remains unchanged after all (the question is scrutinised in the famous Laudan-Worrall debate in the late 1980s; Laudan 1989, Worrall 1988, 1989).

For scholars who agreed that the scientific method is dynamic – like Laudan, Shapere, Toulmin, and Ernan McMullin – the new challenge became demonstrating that the dynamic method thesis did not necessitate a cascade into relativism, or an admission that scientific change had no logic whatsoever. To do so, these scholars posited that an adequate theory of scientific change must broaden its focus and labour to discover the mechanism by which *both* theories *and* methods change (Option 3). Interestingly, however, many of the earliest theories of scientific change which attempted to articulate such a mechanism (such as Laudan's (1984) *reticulated model*) were still meant to be both *descriptive* and *normative*: they attempted to uncover not only how science *does* change, but how it *ought* to change as well. This being said, it became much less obvious how such a theory might have any normative force if individual norms (that is, methods of theory evaluation) changed through time. After all, if methods of theory evaluation do change though time, then the whole enterprise of explicating the one true method of science becomes extremely dubious.

As Figure 1.1 illustrates, however, there is another option available to those who accept the dynamic method thesis: abandoning the search for a general theory of scientific change altogether (Option 2). This does not necessarily imply that changes in scientific theories and methods must be inexplicable. Rather, it opens the door to explaining changes in science (science's theories, methods, organisation, etc.) in terms of processes not unique to scientific communities. Option 2 allows changes in science to be understood as tokens of the sociological, anthropological, or psychological type (but categorically denies that 'science' or 'epistemic changes' are somehow unique and deserving of their own theoretical framework).

I posited that one reason for the unintegration of HS and PS was due to a conflation of two distinct projects: the search for a *descriptive* theory of scientific change and the search for a *normative* theory of scientific change. If the method of science is understood to be common to all scientific communities – the singular 'motor' driving all theory change – then the historical project of *describing* this method simultaneously clarifies the method the current scientific community *ought* to employ: they are the same project.[5] But with the acceptance of the dynamic method thesis, it became increasingly apparent that these projects should remain distinct. The mounting historical case for multiple and changing methods of science made the *normative* philosophical project much more difficult. How, after all, do we discriminate between better or worse methods, or determine the method the scientific community ought to employ today, if methods change with the exigencies of a community's geography, culture, or century? As a result, HS largely dropped the philosophically-informed normative theories they had inherited and, instead, adopted theoretical frameworks that purported to be merely descriptive, such as those provided by sociology and similar human sciences. The general failure of theories offered by scholars pursuing Options 1 and 3 made Option 2 – which is clearly divorced from a philosophy of *science* – a more reasonable choice for many historians of science. John Zammito (2004, p. 120) puts it well:

> [N]o single philosophy of science ... proved adequate to historical purposes ... No such philosophy of science has withstood fundamental [historical] criticism with much of its edifice left standing. That historians have been skeptical and selective with such a volatile fund of critical resources seems eminently sane in that light.

The absence of an historically or philosophically adequate philosophy of science assumed by the choice of Option 2 led historians of science to a break from the normative bent of philosophies of science, and to a shift towards the choice of adequate descriptive theories, like those from sociology, anthropology, and psychology. This makes sense of the exodus of historians of science from PS to either the Sociology of Scientific Knowledge (SSK) or to the pursuit of micronarratives drawing from anthropology and ethnography: both can draw from purely descriptive theoretical frameworks, and are discouraged from imposing normative frameworks.[6]

But the adoption of Option 2 has usually been premised on the supposition that articulations of Option 3 are never able to adequately accommodate the nuances of history, or that they problematically prescribe a normative framework. Neither is necessarily true, however. For instance, despite past failures, it is completely possible for pursuers of Option 3 to craft a solely *descriptive* theory of scientific change that *can* accommodate the nuances of actual historical episodes. Whereas those who subscribe to Option 1, such as Lakatos, might explain the transition from Ptolemaic to Copernican cosmology as a triumph of the (fixed) hypothetico-deductive scientific method,[7] a *scientonomic* theory can show why this transition actually satisfied the Aristotelian-Medieval method employed by the community of the time, while simultaneously insisting that the normativity implied by this conclusion would only have been binding in the European medieval context.

The widespread adoption of Option 2, the abandonment of theories of scientific change, therefore seems to have been based largely on the historical inadequacy of previous attempts to pursue Option 3. Even though these attempts have purportedly articulated mechanisms that explain *both* the changes in theories *and changes in methods*, their normative formulations have not adequately mapped onto documented cases of scientific change in history. But despite the normative heritage of PS, Option 3 allows for attempts at purely *descriptive* theories of scientific change that capture both the changes of theories and methods. Perhaps HS and PS could join forces if a theory could successfully articulate changes in scientific theories and methods in a purely descriptive mode.[8]

1.2 A theory of scientific change: territories, mosaics, patterns

The definition of a particular empirical science is often mapped onto a specific domain or territory in the empirical world. For example, quantum mechanics is defined by limiting its focus to the smallest components of our physical universe and the technological systems that allow us to measure them; biology is defined by its focus on living beings and the physical processes which make life possible, etc. Empirical sciences largely function by observing how objects and processes change within their respective domains, and by producing and assessing theories based on the general patterns discovered within them. In principle, then, it is possible to limit our focus to *epistemic communities*, and to further hone in on the *theories* they *accept* as the *best available descriptions of reality*; in turn, this leads us to the *methods* which they *employ* to assess which theories ought to be considered their *best*.[9] Let us call an epistemic community's web of beliefs and expectations – the set of its accepted theories and employed methods – its *mosaic*.[10] By tracking the changes in a community's mosaic – its accepted theories *and* its employed methods – it is possible to uncover general patterns unique to this domain of reality, should they present themselves.

Indeed, when we thus focus our attention on changes in mosaics, general patterns do emerge despite diverse historical and cultural contexts. It is precisely these patterns in the mosaics of epistemic communities which form the foundations of Hakob Barseghyan's general theory of scientific change. To help introduce this general theory, and the science of science which it helped inspire, let us consider a few such patterns from the history of science itself, beginning with a sketch of the history of medical drug testing: the transition from controlled trials to double-blind trials. This case will help illustrate how methods of theory evaluation change through time in an orderly fashion.

The main goal of the medical drug testing community is to determine whether or not a particular drug is therapeutically effective. Were they to consider whether a certain antidepressant is therapeutically effective or not, therefore, they would be assessing the theoretical claim: 'Antidepressant X is therapeutically effective.' Suppose that a particular community of medical drug testers is well aware of the fact that simply giving a group of depressed people the drug and waiting to see improvement would be an insufficient trial, since they know that improvement (or its absence) can be due to effects unaccounted for by such a study (the diet, relative health, or level of physical activity of the participants, for instance). This community therefore tests the theory 'Antidepressant X is therapeutically effective' by dividing its trial participants into a control group, which does not receive Antidepressant X, and the variable group, which does. If the variable group's improvement is markedly better than that of the control group, the community can accept the theory about the therapeutic efficacy of Antidepressant X.

Suppose, however, that this community then learns about the *placebo effect*. The question then arises: when testing the drug's next iteration (Antidepressant Y) and after having learned about the placebo effect, would a controlled trial now suffice to satisfy the expectations of the community that the theory 'Antidepressant Y is therapeutically effective' is the best available theory? No. Upon learning about and accepting the possibility of the placebo effect to influence assessments of the drug's efficacy, the community would likely devise a means of forestalling or minimising the impact of the placebo effect. In this case, let us assume that the community implements this new expectation by devising the single-blind trial, veiling from participants whether they are part of the variable or control group. Should Antidepressant Y prove therapeutically effective after single-blind trials, the community would accept the theory about Antidepressant Y's efficacy.

Suppose that this community then learns about and accepts the reality of *experimenter's bias* before testing the drug's next iteration, Antidepressant Z. Certainly a single-blind trial would no longer be sufficient to accept the drug's efficacy. The community devises a double-blind trial to mask the identity of the variable and control groups from both those administering the drugs and the trial participants, thus forestalling the influence of experimenter's bias. Should Antidepressant Z prove therapeutically effective after double-blind trials, the community could accept the theory about the drug's efficacy.

While obviously idealised, this brief thought experiment tracks onto the actual history of drug trial testing and helps to demonstrate how the expectations of this community change when they accept new theories (Figure 1.2).[11] In fact, it seems that the relationship between new methods and new theories is a *logical* one.

Framing this insight in terms of the community's mosaic, therefore, we can articulate the first general pattern of scientific change regarding the employment of new methods:

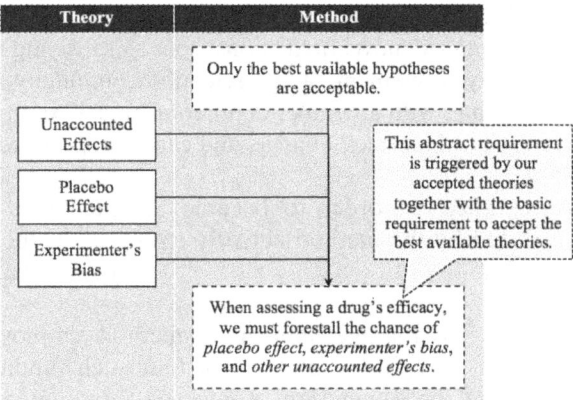

Figure 1.2 Relationship between new methods and new theories

> Method Employment: A method becomes employed only when it is deducible from some subset of other employed methods and accepted theories of the time.[12]

This mechanism seems to capture changes in methods on any scale and in any period of history, from the relatively minute changes in requirements illustrated in the history of drug trial testing, to the employment of the hypothetico-deductive (HD) method at the dawn of the scientific revolution (Barseghyan 2015, pp. 133, 143, 148). Indeed, the HD method can be shown to be a deductive consequence of theories in Cartesian and Newtonian science: if all natural phenomena are ultimately caused by the complex interactions of matter in ways that are not immediately apparent or intuitive, then without direct observation of these mechanisms it is in principle possible to posit an infinite number of post-hoc explanations for phenomena. The new HD method incorporated these assumptions, and its requirement for confirmed novel predictions helped curtail post-hoc explanations.

Considering how René Descartes' natural philosophy became accepted by the scientific community will help illustrate another regularity in mosaics' change. In his *Discourse on Method* and *Meditations on First Philosophy*, Descartes (1637, 1641) builds his natural philosophy and his approach to the material world from a foundation of intuitively graspable truths, such as his famous *cogito* and his claim that the only essential feature of matter is *extension*. It was precisely due to Descartes' presentation of his science as a system of deductive consequences of clear and distinct truths that his science satisfied the Aristotelian-Medieval method of the scientific community of the time. For the Aristotelian-Medieval community, a theory was acceptable if it was intuitively graspable by an experienced person, or if it followed logically from other accepted intuitive truths (Barseghyan 2015, pp. 143–145).

Descartes' science became accepted on the Continent – not due to its empirical success or by confirming any novel predictions – but by satisfying the expectations and criteria for theory assessment of the scientific community of his time. As Barseghyan demonstrates with a number of other historical examples,[13] this is one instance of a regularity in mosaics' change, and can be articulated as follows:

> Theory Acceptance: In order to become accepted into the mosaic, a theory is assessed by the method actually employed at the time.
>
> (Barseghyan 2015, p. 129)

These two mechanisms, the mechanism of method employment and the mechanism of theory acceptance, are two of four such fundamental regularities which constitute the foundation of a general *descriptive* theory of science: the four laws of scientific change (Figure 1.3) (ibid., p. 123).

As a general descriptive theory of scientific change, these laws should be applicable to any epistemic community's mosaic in any culture, at any time. The choice of the term 'law' is not meant to suggest that these specific formulations are immune to correction, nor does it commit those who accept the theory to a robust understanding of the ontological status of said 'laws' (this remains an open question for the theory). But 'law' does evoke a sense of transhistorical permanence and axiomatic systematicity, and intentionally so: the community of scholars working on this theory have derived 23 additional theorems from the laws, which have been used to discover and to understand other features of the process of scientific change.[14]

For instance, the sociocultural factors theorem contributes to a long debate in HPS as to whether sociocultural factors can (or ought) to be part of theory assessment. Derived from the second law, the *sociocultural factors theorem* states that these factors *can* affect theory assessment *provided they are part of the method of the time*. This does not rule out the influence of sociocultural factors *a priori*, as the positivists had, but rather makes their impact an empirical question for historians of science.[15] This historical sensitivity is also showcased in the *contextual appraisal theorem*, which was derived from the

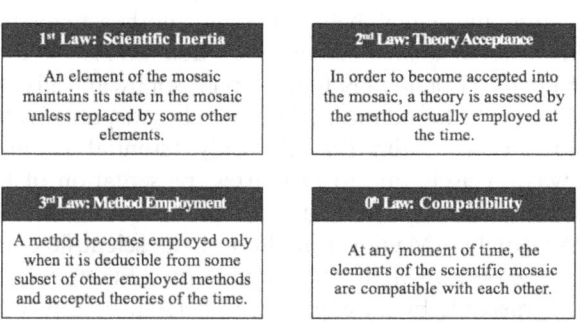

1st Law: Scientific Inertia	2nd Law: Theory Acceptance
An element of the mosaic maintains its state in the mosaic unless replaced by some other elements.	In order to become accepted into the mosaic, a theory is assessed by the method actually employed at the time.
3rd Law: Method Employment	0th Law: Compatibility
A method becomes employed only when it is deducible from some subset of other employed methods and accepted theories of the time.	At any moment of time, the elements of the scientific mosaic are compatible with each other.

Figure 1.3 Four laws of scientific change

first and second laws of scientific change. The theorem emphasises that theories are only accepted or rejected in a specific mosaic following an assessment by the employed method *of that mosaic* (not by a method or set of requirements foreign to it). Simply, theories are assessed by that community's norms, not our contemporary ones.[16] We could avoid the tired anachronistic retelling of the Galileo affair (the freethinking scientist persecuted by the dogmatic anti-science establishment; Dreger 2015, pp. 11–18) if we appreciated the lesson of the contextual appraisal theorem: Galileo's case for a heliocentric cosmology *did not appeal to the method employed in sixteenth century Italy*, namely, the Aristotelian-Medieval method of intuitive truth. Galileo was not punished for being scientific, he was punished for *not being scientific enough*.

1.3 Scientonomy: integrating HS & PS

By focusing on the domain of communities' mosaics and describing the regularities found therein, the theoretical framework developed around the laws of scientific change lays the groundwork for an empirical science of science, relying heavily on the history of science and effectively naturalising a philosophy of science. The idea of transforming HPS into an empirical science of science has been suggested before, but there are few examples of it in practice.[17] The community of scholars that has committed to using Barseghyan's theory in historical investigations and developing the theory further has dubbed their science of science *scientonomy*.

The clearest way that scientonomy represents a reintegration of PS and HS is its explicit dependence upon both *theoretical scientonomy* and *observational scientonomy*. Like traditional PS, theoretical scientonomy seeks to identify and articulate the mechanisms which govern scientific change. Unlike traditional PS, however, theoretical scientonomy does not seek to answer the normative question of how science *ought* to proceed.[18] (Theories about communities' norms are not necessarily normative theories.) Observational scientonomy has the entire past and present of science and scientific change as its domain, much like today's HS. Whereas contemporary HS either adopts theoretical frameworks from other domains like sociology or anthropology, or seeks to avoid overarching theoretical metanarratives altogether, observational scientonomy explicitly adopts scientonomy as its guiding theory.[19] While scientonomists might individually be better suited for either observational or theoretical scientonomy, observational and theoretical scientonomy are complementary and mutually imbricated, constituting the diastole and systole of the field's life.[20] This of course mirrors the mutual dependence between the theoretical and observational/experimental dimensions of other sciences, and especially the necessary synthesis of idiographic and nomothetic in historical sciences like palaeontology (Turner 2014).

There is a sense in which the integration between PS and HS in scientonomy is the result of a conscious embrace of what the philosophy of science has typically framed as a *problem*: the theory-ladenness of observations.

There is an explicit recognition that one cannot approach the history of science without a theory (even if unconsciously). And there's an explicit recognition that this theory, being the product of fallible investigators, can always be improved and revised based on new information or approaches. Working together in the scientonomy community, historians and philosophers, observational and theoretical scientonomists, have taken this general pragmatism and commitment to piecemeal improvement and operationalised it at the University of Toronto's Institute for the History and Philosophy of Science and Technology. We will now turn our attention to the scientonomy workflow and the present and future opportunities produced by this version of an iHPS.

1.3.1 The scientonomy workflow

The scientonomy workflow is an iterative one, with each stage designed to contribute to the next, and with no proper point of entry (Figure 1.4). That being said, the *seminar* is the oldest component, and its discussions led to the implementation of the remainder of the workflow, so it will be my point of entry.

The Seminar: The seminar introduces participants to the history of theories of scientific change and familiarises them with the theoretical foundations of scientonomy, all with the goal of producing a new list of open questions for theoretical and observational scientonomists to tackle. The seminar has run once yearly since 2013, and has produced 69 open questions to date. Open questions are then publicly listed on the online *Encyclopedia of Scientonomy.*[21]

The Journal:[22] *Scientonomy: Journal for the Science of Science* is an open-access, peer-reviewed journal established in 2016 to 'seek solutions to topical issues in theoretical scientonomy; apply the currently accepted theory to historical cases of scientific change' and 'engage in a philosophical analysis of the key concepts of scientonomy.'[23] Scholars interested in addressing any of the

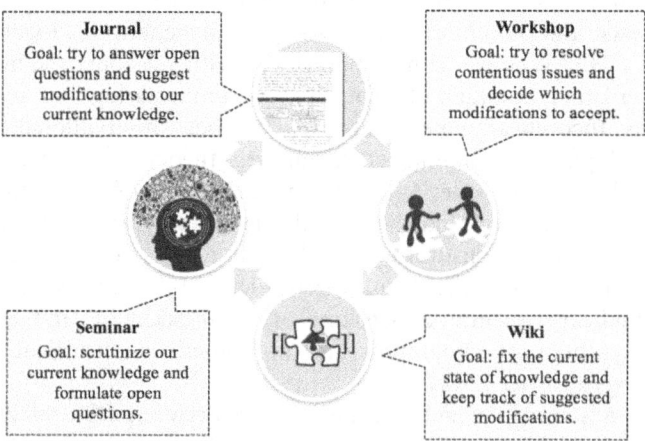

Figure 1.4 Scientonomy workflow

open questions raised in seminars – or any other relevant topic – can submit their papers for review. In the interest of contributing positively to the development of scientonomic knowledge, each paper is required to include a list of suggested modifications for the community of scientonomy to assess. The modifications of published papers are then posted to the online *Encyclopedia of Scientonomy* to allow for public consideration and communal discussion.

The Workshop: Should any suggested modification discussed by the community ever become contentious or in need of focused debate and discussion, there is the possibility of holding a workshop. Unlike typical academic conferences, where goals range from general presentations of research to networking, the workshop would be designed with the explicit intention of reaching a consensus regarding the problematic suggested modification. The scientonomy community has not yet had the need for a workshop.

The Encyclopedia (Wiki):[24] The online *Encyclopedia of Scientonomy* is an easily accessible means of ascertaining the current state of scientonomic knowledge. As noted above, contributions from every stage of the workflow are regularly submitted to the *Encyclopedia* to serve this goal. Notably, the *Encyclopedia* has been organised and coded in a *semantically meaningful* way, facilitating data entry, data organisation, and data presentation.[25] In many ways it is the most critical piece of infrastructure for the functioning of the current workflow. This foray into web design and database management, and the questions it has raised for how scientonomy itself is organised and presented, has also helped to inspire the Tree of Knowledge Project, which I will discuss below.

The iterative scientonomic workflow is meant to facilitate the study of scientific change by encouraging a piecemeal, transparent, accessible process that emphasises consensus building and problem solving. There should never be a question as to where the field stands with respect to a certain question, theory, or suggested modification, and the input and expertise of theoretical and observational scientonomists can make positive contributions at any stage of the workflow. Since the establishment of the *Journal* in 2016, 23 modifications have been suggested, and eight have been accepted by community consensus via the *Encyclopedia*.

1.3.2 The workflow: success and promise

It is worth considering a brief concrete example of the workflow in action. Intrigued by questions raised in the 2016 seminar regarding authority delegation between scientific communities,[26] Mirka Loiselle submitted a paper to *Scientonomy* investigating the authority delegation structures between epistemic communities interested in the authenticity of the artworks of Amedeo Modigliani and Pierre-Auguste Renoir. In the case of Modigliani, Loiselle found that while the art market community had delegated authority to the catalogue raisonné compiled by expert Ambrogio Ceroni as the principal means of determining the authenticity of Modigliani's work, they had also evidently delegated authority to the catalogue raisonné compiled by Marc Restellini in cases of paintings not found in Ceroni's.[27] In the case of the authenticity of Renoir's works, Loiselle discovers a similar structure of multiple, hierarchical authority delegation, with the art market community delegating authority first to the Wildenstein Institute and then to the Bernheim-Jeune Gallery. These studies in the history of art authentication led Loiselle to suggest new models of authority delegation in addition to the results of her historical investigation – led explicitly by the theoretical framework of scientonomy. After peer review, the paper was accepted and published in *Scientonomy*, and the merits of the suggested modifications (to accept the aforementioned types of authority delegation) are currently being discussed on the *Encyclopedia*.[28]

The relative successes of the online components of the workflow, namely the *Encyclopedia* and the *Journal*, inspired a vision for a larger, bolder, database-driven project provisionally named the *Tree of Knowledge Project*. Along with its system of laws, scientonomy has developed a robust taxonomy/ontology (with some help from the *Encyclopedia* project) with which it can describe a diversity of epistemic stances taken by different epistemic agents towards different epistemic elements.[29] The combination of scientonomic theory and scientonomic taxonomy can allow for the building of an online database of epistemic communities, their theories, and their methods. Similar to other massive online database projects like the Open Tree of Life,[30] the SIMBAD Astronomical Database,[31] and the Database of Religious History,[32] the Tree of Knowledge could allow users to survey the mosaics of epistemic communities across the globe and through history. Historians and philosophers of science would be some of the best equipped to undertake this project, and there are few scholarly communities it could potentially serve better than HPS. But as anyone who has ever designed a database knows, a shared taxonomy is necessary: there must be agreed-upon definitions for terms like epistemic agent, theory, method, acceptance, etc. Because scientonomy's taxonomy derives from an historically nuanced, non-normative theory of scientific change, scientonomy seems poised to take the lead on such a project.[33]

Conclusion

In this chapter I have demonstrated that a way in which HS and PS can be fruitfully united into one integrated discipline is by reconceiving them as the observational and theoretical components of a single empirical *science of science*. By outlining Barseghyan's theoretical basis for scientonomy and introducing the operationalised workflow of the burgeoning community of *scientonomists* at the University of Toronto, I have also provided a proof of concept of this kind of iHPS, and have suggested future opportunities for its growth.

It certainly is not the case that scientonomy is the *only* way of integrating HPS, nor does it purport to replace all other PS and HS. For instance, as a descriptive enterprise scientonomy is silent on normativity in scientific change (how *should* science change?), and it also does not engage with the fundamental metaphysics of science itself, both of which are still comfortably within the realm of PS (though these investigations should be historically informed). With respect to HS, since scientonomy's theory mainly focuses on *communities'* theory appraisal and method employment, its historical interests cannot (yet) extend much beyond them. There is still great value in historically investigating the biographies of individual scientists, scientific institutions, micronarratives, material cultures, theory creation, and laboratory cultures, for instance, provided one is conscious of one's theoretical commitments, their strengths, and their weaknesses.

In 2016 Larry Laudan penned another retrospective piece in *Studies in the History and Philosophy of Science*, this time with his wife, Rachel. In it, they reflect on the production and (relatively poor) reception of their *Scrutinizing Science: Empirical studies of scientific change*, but also comment on the potential for a re-emergence of an empirical iHPS. They conclude that they

> cannot help speculating what would happen if historians and philosophers of science followed their scientific colleagues from time to time, pooling expertise to work on a common problem. And we remain convinced that to flourish the meta-study of science must be self-reflexive, as ready to test its own claims as scientists are to test theirs.[34]

As the scientonomy community grows, perhaps they will not have to speculate much longer.

Notes

1 Jutta Schickore (2011, pp. 453–454) has compiled an excellent list of these contributions to integrated HPS. Additionally, communities such as &HPS, the Committee for Integrated HPS, and the scientonomy community indicate a growing interest in iHPS. This volume, of course, is also the result of the postgraduate forum on 'The Past, Present, and Future of Integrated HPS' that took place at the University of Leeds in January 2017.

2 As I will illustrate in Section 1.1, HPS is a term adopted by historians and philosophers of science to unify their different scholarly approaches to science under one administrative roof, which I sometimes call the 'field of HPS'. A historian of science can theoretically ignore the methods, concerns, and lessons of the philosophy of science, and a philosopher of science can do the same for the history of science, and – working independently – they are still very much members of the field of HPS. iHPS, however, requires an intentional attempt to meld these two approaches to science, bringing the methods, concerns, and lessons of philosophy of science to bear on the history of science, and vice versa. So while all iHPS is HPS, not all HPS is (or needs to be) iHPS.

3 Two classic examples of this unfavourable reception of Lakatos' rational reconstructions among historians are Stephen Shapin (1982), and Paul Feyerabend (1975).

4 See Whewell (1949); A. Rupert Hall ([1954] 1966).

5 See point 1 in section 1.1.1, above (p. XX).

6 See Bloor's riposte to Laudan on the topic of 'normative rationality' in David Bloor (1981), pp. 207–211; also, see Bruno Latour and Steve Woolgar (2013),

especially pp. 32–33. Other corroborating narratives include Gerd Buchdal (1965), and Lynn Nyhart (2016).

7   See Imre Lakatos and Elie Zahar (1978).

8   This descriptive project of course does not rule out the possibility that HS and PS can be fruitfully integrated in a way that is explicitly normative. For an excellent example of a Popperian, normative, historically oriented iHPS today, see Eugenio Petrovich's Chapter 4 in this volume.

9   While the term *epistemic community* is derived from the pioneering works of Karin Knorr-Cetina (1982; 1999), the working definition in scientonomy is simply 'a community that has a collective intentionality to know the world.' Cf. N. Overgaard (2017).

10 Epistemic communities and mosaics, like Ludwig Fleck's *Denkkollektiv* and *Denkstil*, demonstrate a shared commitment to studying science as a social phenomenon. However, scientonomy's clear delineation between theories and methods – and the lawlike dynamics these elements historically display when so-delineated – sets this approach apart from those of Fleck or Thomas Kuhn.

11 It is important to note that in Kaptchuk (1998) and Justman (2010) the shift to blinded or masked trials is always preceded by the acceptance of the placebo effect's capacity to distort results.

12 'Mechanism of Method Employment', in *Encyclopedia of Scientonomy*, https://www.scientowiki.com/Mechanism_of_Method_Employment (accessed 15.04.2018).

13 Some examples from Barseghyan's *The Laws of Scientific Change* include: the acceptance of Einstein's theory of general relativity (p. 130); the acceptance of Newtonian natural philosophy (p. 130); the rejection of Galilean heliocentrism (p. 187); the acceptance of Cartesian natural philosophy in Paris despite the acceptance of the Catholic doctrine of the Real Presence (pp. 191–195).

14 The current list of derived theorems can be found here: 'Theory of Scientific Change: Theorems', in *Encyclopedia of Scientonomy*, https://www.scientowiki.com/The_Theory_of_Scientific_Change#Theorems (accessed 15.04.2018).

15 See Lakatos (1978), pp. 102, 114, 118–121.

16 In Chapter 6 of this volume Massimiliano Simons marvelously describes the French historical epistemological tradition, whose understanding of scientific communities' 'obligations' (in Isabelle Stengers' sense, and opposed to 'requirements') seems to line up quite well with what scientonomists mean by an *employed method*.

17 See Donovan, Laudan and Laudan (1988), and compare with the Laudans' retrospective in Larry Laudan and Rachel Laudan (2016). Ronald N. Giere (2012, pp. 60–61) has recently suggested a similar move towards naturalising the philosophy of science and empirically testing it.

18 'Scientonomy', in *Encyclopedia of Scientonomy*, https://www.scientowiki.com/Scientonomy (accessed 15.04.2018).

19 Ibid.

20 Cf. Giere 2012, pp. 61–62.

21 'List of Open Questions', in *Encyclopedia of Scientonomy*, https://www.scientowiki.com/List_of_Open_Questions (accessed 31.05.2018).

22 *Scientonomy: Journal for the Science of Science*, www.scientojournal.com (accessed 15.04.2018).

23 'Journal of Scientonomy', in *Encyclopedia of Scientonomy* (25.6.2017), https://www.scientowiki.com/Journal_of_Scientonomy)accessed 15.04.2018).

24 *Encyclopedia of Scientonomy*, https://www.scientowiki.com (accessed 15.04.2018).

25 Ibid.

26 The definition of *authority delegation* in scientonomy is the following: 'Community A is said to be delegating authority over topic *x* to community B *iff* (1) community A accepts that community B is an expert on topic *x* and (2) community A will accept a theory on topic *x* if community B says so.' This was first suggested in Overgaard and Loiselle (2017), 11–18.

27 A catalogue raisonné is a document which lists all known, authenticated works of a particular artist, and is used as an authoritative document by both art experts and those interested in purchasing or selling artworks (the art market community). These works are authenticated using forensic dating techniques, the works' provenance, and the connoisseurship of experts, family, and friends. Mirka Loiselle (2017), p. 42.

28 The modifications suggested in this paper can be found on the *Encyclopedia*: 'Loiselle 2017', in *Encyclopedia of Scientonomy*, https://www.scientowiki.com/Loiselle_(2017) (accessed 9.7.2018).

29 'Ontology of Scientific Change', in *Encyclopedia of Scientonomy*, https://www.scientowiki.com/Ontology_of_Scientific_Change (accessed 15.04.2018).

30 *Open Tree of Life*, https://tree.opentreeoflife.org/ (accessed 30.11.2017).

31 *SIMBAD Astronomical Database - CDS (Strasbourg)*, http://simbad.u-strasbg.fr/simbad/ (accessed 30.11.2017).

32 *The Database of Religious History*, https://religiondatabase.org/landing/ (accessed 15.04.2018).

33 'Tree of Knowledge Project', in *Encyclopedia of Scientonomy*, https://www.scientowiki.com/Tree_of_Knowledge_Project (accessed 15.04.2018).

34 Laudan and Laudan (2016), p. 77.

Acknowledgements

The author would like to thank Hakob Barseghyan and the editors of this volume for their substantial guidance on previous drafts. The author would also like to acknowledge the financial support of the Institute for the History and Philosophy of Science and Technology at the University of Toronto and the Ontario Graduate Scholarship.

Bibliography

Barseghyan, Hakob, *The Laws of Scientific Change* (Cham: Springer, 2015).

Bloor, David, 'The Strengths of the Strong Programme', *Philosophy of the Social Sciences*, 11(1981), 199–213.

Buchdal, Gerd, 'A Revolution in Historiography of Science', *History of Science*, 4 (1965), 55–69.

Butterfield, Herbert, *The Origins of Modern Science: 1300–1800* (London: G. Bell and Sons Ltd, 1949).

Caneva, Kenneth, 'What in Truth Divides Historians and Philosophers of Science?', in *Integrating History and Philosophy of Science*, ed. by S. Mauskopf and T. Schmaltz (Dordrecht: Springer, 2012), 49–56.

Descartes, René, *Discourse on Method* (1637).

Descartes, René, *Meditations on First Philosophy* (1641).

Donovan, Arthur, Larry Laudan, and Rachel Laudan, eds., *Scrutinizing Science: Empirical Studies of Scientific Change* (Dordrecht: Kluwer Academic Publishers, 1988).

Dreger, Alice, *Galileo's Middle Finger* (New York, NY: Penguin Press, 2015).

Encyclopedia of Scientonomy, www.scientowiki.com (accessed 15. 4. 2018).

Feyerabend, Paul, *Against Method* (London: New Left Books, 1975).

Giere, Ronald N., 'History and Philosophy of Science: Intimate Relationship or Marriage of Convenience?', *The British Journal for the Philosophy of Science*, 24. 3(1973), 282–297.

Giere, Ronald N., 'History and Philosophy of Science: Thirty-Five Years Later', in *Integrating History and Philosophy of Science*, ed. by S. Mauskopf and T. Schmaltz (Dordrecht: Springer, 2012), 59–65.

Hall, A. Rupert, *The Scientific Revolution, 1500–1800: The Formation of the Modern Scientific Attitude* (Boston, MA: Beacon Press, 1966).

'Journal of Scientonomy', in *Encyclopedia of Scientonomy*, http://www.scientowiki. com/Journal_of_Scientonomy (accessed 15. 4. 2018).

Justman, S., 'Imagination's Trickery: The Discovery of the Placebo Effect', *Journal of the Historical Society*, 10(2010), 57–73.

Kaptchuk, T. J., 'Intentional Ignorance: A History of Blind Assessment and Placebo Controls in Medicine', *Bulletin of the History of Medicine*, 72(1998), 389–433.

Knorr-Cetina, Karin, *Epistemic Cultures: How the Sciences Make Knowledge* (Cambridge, MA: Harvard University Press, 1999).

Knorr-Cetina, Karin, 'Scientific Communities or Transepistemic Arenas of Research?: A Critique of Quasi-Economic Models of Science', *Social Studies of Science*, 12(1982), 101–130.

Kuhn, Thomas, 'The Relations Between History and the Philos Sci,' in *The Essential Tension: Selected Studies in Scientific Tradition and Change* (Chicago, IL: University of Chicago Press, 1977), 3–30.

Lakatos, Imre, 'History of Science and its Rational Reconstruction', in *Philosophical Papers*, I, (Cambridge: Cambridge University Press, 1978), 102–138.

Lakatos, Imre and Elie Zahar, 'Why Did Copernicus's Research Programme Supersede Ptolemy's?', in *Philosophical Papers*, I, (Cambridge: Cambridge University Press, 1978), 168–192.

Laudan, Larry, *Science and Values* (Berkeley, CA: University of California Press, 1984).

Laudan, Larry, 'If It Ain't Broke Don't Fix It', *British Journal for the Philosophy of Science*, 40(1989a), 369–375.

Laudan, Larry, 'Thoughts on HPS: 20 Years Later', *Studies in History and Philosophy of Science*, A.20(1989b), 9–13.

Laudan, Larry and Rachel Laudan, 'The Re-Emergence of Hyphenated History-and-Philosophy-of-Science and the Testing of Theories of Scientific Change', *Studies in History and Philosophy of Science, A*, 59(2016), 74–77.

Latour, Bruno and Steve Woolgar, *Laboratory Life: The Construction of Scientific Facts* (Princeton, NJ: Princeton University Press, 2013).

'List of Open Questions', in *Encyclopedia of Scientonomy*, http://www.scientowiki.com/ List_of_Open_Questions (accessed 15. 4. 2018).

Loiselle, Mirka, 'Multiple Authority Delegation in Art Authentication', *Scientonomy: Journal for the Science of Science*, 1(2017), 41–53.

'Mechanism of Method Employment', in *Encyclopedia of Scientonomy*, http://www. scientowiki.com/Mechanism_of_Method_Employment (accessed 15. 4. 2018).

Nola, Robert, and Howard Sankey, eds., *After Popper, Kuhn and Feyerabend. Recent Issues in Theories of Scientific Method* (Boston, MA: Kluwer Academic Publishers, 2000).

Nyhart, L. K., 'Historiography of the History of Science', in *Companion to the History of Science*, ed. B. Lightman (Chichester, West Sussex: Wiley Blackwell, 2016), 6–22.

'Ontology of Scientific Change', in *Encyclopedia of Scientonomy*, https://www.sciento wiki.com/Ontology_of_Scientific_Change (accessed 15. 4. 2018).

Open Tree of Life, https://tree.opentreeoflife.org/ (accessed 15. 4. 2018).

Overgaard, Nicholas, 'A Taxonomy for the Social Agents of Scientific Change', *Scientonomy: Journal for the Science of Science*, 1(2017), 55–62.

Overgaard, Nicholas, and Mirka Loiselle, 'Authority Delegation', *Scientonomy: Journal for the Science of Science*, 1(2017), 11–18.

Schickore, Jutta, 'More Thoughts on HPS: Another 20 Years Later,' *Perspectives on Science*, 19. 4(2011), 453–481.

'Scientonomy', in *Encyclopedia of Scientonomy*, http://www.scientowiki.com/Sciento nomy (accessed 15. 4. 2018).

Scientonomy: Journal for the Science of Science, www.scientojournal.com (accessed 15. 4. 2018).

Shapere, Dudley, 'The Character of Scientific Change,' in *Scientific Discovery, Logic and Rationality*, ed. T. Nickles (Dordrecht: Reidel, 1980), 61–101.

Shapin, Stephen, 'History of Science and its Sociological Reconstructions', *History of Science*, 20. 3(1982), 157–211.

SIMBAD Astronomical Database - CDS (Strasbourg), http://simbad.u-strasbg.fr/sim bad/ (accessed 15. 4. 2018).

The Database of Religious History, https://religiondatabase.org/landing/ (accessed 15. 4. 2018).

'Theory of Scientific Change: Theorems', in *Encyclopedia of Scientonomy*, https://www.sci entowiki.com/The_Theory_of_Scientific_Change#Theorems (accessed 15. 4. 2018).

'Tree of Knowledge Project', in *Encyclopedia of Scientonomy*, http://www.scientowiki. com/Tree_of_Knowledge_Project (accessed 15. 4. 2018).

Turner, Derek D., 'Philosophical Issues in Recent Paleontology', *Philosophy Compass*, 9. 7(2014), 494–505.

Whewell, William, *The Philosophy of the Inductive Sciences, Founded upon their History*, 2 vols (London: John W. Parker, 1840).

Worrall, John, 'Fix it and be Damned: A Reply to Laudan', *British Journal for the Philosophy of Science*, 40(1989), 376–388.

Worrall, John, 'Review: The Value of a Fixed Methodology', *British Journal for the Philosophy of Science*, 39(1988), 263–275.

Zammito, John H., *A Nice Derangement of Epistemes* (Chicago, IL: University of Chicago Press, 2004).

2 Understanding past research practice: A case for iHPS

Caterina Schürch

Introduction

Can the disciplines of History of Science and Philosophy of Science be successfully integrated and thrive together? Or would it be better for scholars with an interest in advancing our understanding of science to adhere to one discipline and pursue either historical analyses or philosophical inquiries? As the ongoing debate about Integrated History and Philosophy of Science (iHPS) proves, there is strong support for both arguments. On the one side there is the anti-integration camp: scholars who insist that the two disciplines be kept distinct and who regard iHPS as a 'largely unsuccessful experiment' or even a 'failure' (Shapin and Schaffer 2011, p. xxi; Kuukkanen 2016). Pro-integration supporters, however, believe that history can help provide insights into philosophical problems, while scholarship in philosophy can improve historiography. An early example of pro-integration thinking is an anonymous mid-1940s memorandum of Cambridge University's Faculty Board of 'Biology B', which states that 'the importance of History of Science lies not in learning the sequence of discovered fact, but in appreciating the evolution of Natural Philosophy and Scientific Method'.[1]

In the past, the iHPS debate has often focused on general statements such as that good history needs philosophy or that historiography can never result in good philosophy.[2] Rather than follow suit, I have turned the problem upside down and asked the following question: Are there metascientific issues that would benefit from being examined in both historical and philosophical contexts? In this chapter I argue that the problem of understanding past scientific research practice – that is, the particular way in which scientists processed a given research problem – is such an issue. In addition, I show that integrating concepts and methods from History of Science and Philosophy of Science is not a methodological crime and that integrating the two approaches advances metascience.

In the first section of this chapter, I claim that integrating historical and philosophical analyses is an effective and legitimate way of solving research problems that fulfil the following criteria: firstly, they address questions pertaining to History of Science as well as to Philosophy of Science;

secondly, the two questions are interdependent, such that one cannot be answered unless the other is also answered. The real question is whether there are such metascientific research problems – particularly as critics of iHPS maintain that philosophers should not be allowed to write history and that historians cannot further philosophy. I argue that these difficulties are surmountable and demonstrate in the remainder of the chapter that iHPS studies are both feasible and desirable. In section 2, I maintain that the issue of understanding past research practice constitutes a suitable research problem for iHPS, as it fulfils the criterion introduced in section 1: to explain scientists' research practices, material, social and epistemic factors need to be considered together, which is why the input from both History of Science and Philosophy of Science is necessary. In section 3, I illustrate these claims with reference to philosophical work on the search for mechanisms in biology as well as historians' accounts of research between the two world wars in the field of physico-chemical biology. I argue that mainstream History of Science and mainstream Philosophy of Science tend to leave interesting questions unanswered – questions that iHPS scholarship can resolve. This bodes well for its future: philosophy offers a coherent set of epistemic goals and norms that can potentially explain how scientists in the past conceptualised and solved their research problems; historiography, on the other hand, can reveal whether or not the norms and goals discussed in philosophy were indeed pertinent to scientific practice. Finally, in section 4, I demonstrate how past research practice can be investigated in a manner that is both methodologically sound and relevant to historians and philosophers of science.

2.1 The general structure of suitable research problems for iHPS

The Committee for Integrated HPS stated, on the occasion of the group's first &HPS workshop in Pittsburgh in 2007, that iHPS 'is work that is both historical and philosophical' and that the shared goal of understanding science 'can be pursued by dual, interdependent means'.[3] This section focuses on the general structure of research problems that are particularly pertinent to iHPS studies, with a view to explaining the concept of 'interdependent means'. My argument rests on the idea that, by definition, iHPS is the method of choice for solving metascientific research problems that require historical as well as philosophical analyses. Since this implies that iHPS involves interdisciplinary work, I have used concepts of cross-disciplinary collaboration in order to specify the characteristics of metascientific research problems suited for iHPS.

2.1.1 Interrelated goals – the rationale behind integration

Collaboration is commonly taken to be a consequence of goal interdependence.[4] As social psychologist Morton Deutsch puts it, people cooperate if their respective goals are positively interdependent. 'If you're positively linked with another', Deutsch (2006, p. 24) explains, 'then you sink or swim together'.

Admittedly, iHPS studies need not be the result of collaborative work between historians and philosophers: they may also be the work of individual scholars educated in the two fields and skilled in both methods. In either case, however, the objective of iHPS is to respond to historical as well as to philosophical concerns. I argue that this is reflected in the structure of research problems suited to iHPS as they draw attention not only to issues typically addressed in History of Science but also to questions discussed in Philosophy of Science. In the spirit of Deutsch's notion of cooperation, the two issues tackled in an iHPS study are interdependent in that we cannot answer one question without answering the other: *They are linked in such a way that the probability of answering the question pertaining to history is positively correlated with the probability of resolving the philosophical issue.* Integrating historical and philosophical methods to solve such a research problem is, therefore, not only effective but also essential. One may, however, voice doubts about whether there are in fact research problems that really address both historiographical and philosophical questions. In other words: Are there questions asked in History of Science that can only be answered appropriately with the input of philosophy? And are there philosophical questions that would benefit from historical analyses?

Many scholars are highly sceptical about this line of thought: history and philosophy are separate disciplines, they argue, with different metaphysical commitments, goals, norms and methods.[5] For them, iHPS is, at best, a waste of time. They believe that the two disciplines are irrelevant to each other in that historical analyses cannot effectively help us to answer philosophical questions, and vice versa. Some critics also maintain that integrating the two approaches hampers both philosophical and historical analyses, and thus impedes the overarching goal of understanding science.[6] The following is a brief survey of these anti-integration arguments.

2.1.2 Doubts about the value of History of Science to Philosophy of Science

Let us begin with the argument that History of Science is irrelevant to Philosophy of Science. Philosopher Joseph Pitt (2001, p. 373) holds that 'even very good [historical] case studies do no philosophical work'. According to Pitt, philosophers 'seek universals' and are interested in the normative, that is, they aim to show what ought to be meant by x or y (ibid., p. 374). Philosophy's goals cannot be achieved by studying history. Norwood Russell Hanson (1962, p. 581) has made a similar point, characterising Philosophy of Science as 'an analysis and an appraisal of the rationale and logical justification of scientists doing and saying what they do'. It moreover looks as if philosophers are disinclined to limit their accounts of science to what in fact happened in the history of science. Cassandra Pinnick and George Gale (2000, p. 114) point out that 'philosophers allow virtually no empirical constraints to limit their free construction of examples'. So how could History of Science ever be considered relevant to Philosophy of Science?

Critics further warn that scanning history for evidence to support philosophical claims ends up ruining philosophy: it encourages us to believe that a given philosophical account is supported or rejected by historical data. This reasoning is, however, severely hindered by the problem of induction. Even if we were to gather data from many different historical cases, we would not be in a position to draw general conclusions from them.[7] Moreover, it is highly likely that scholars select historical cases that support their own philosophical interests and viewpoints. Thomas Nickles (1995, p. 141) compares the practice of undertaking case studies with Bible reading: 'If one looks long and hard enough, one can find an isolated instance that confirms or disconfirms almost any claim.' Hence, iHPS studies reveal little about science but everything about our own preconceptions.[8] Taken together, critics do not see how the results of historians can have an effect on philosophy as philosophical claims should not and cannot be based on historical data.[9]

2.1.3 Doubts about the value of Philosophy of Science to History of Science

The main drawback of selection bias applies to historiography as well, for various philosophical accounts can be used to analyse history. Different philosophical accounts emphasise different aspects of science, so the resulting descriptions of historical events will diverge accordingly.[10] How does one decide between disparate, often conflicting historical accounts? According to Nickles (1995, pp. 140–41), historicists dismiss philosophers' overall conception of scientific practice as 'wrongheaded, indeed, well-nigh theological, otherworldly, and transcendent', because philosophers constantly refer to reason, rationality and truth. However, historians, states Pitt (2001, p. 376), attempt to reveal 'the social and intellectual factors that might be said to motivate the views expressed by the particular historical figure under consideration'. Indeed, the historian L. Pearce Williams (1975, p. 253) has claimed that history written by philosophers resulted in 'falsehood and bad scholarship'. According to both Pitt and Williams, imposing modern philosophical ideas on history distorts our reading of historical sources. Not without reason, therefore, are history students 'routinely admonished to avoid anachronism at all costs' (Jardine 2000, p. 251).

But why would one even come up with the idea of introducing Philosophy of Science to the study of History of Science? Historian Clayton Roberts (1996, p. 8) states that 'historical knowledge concerns the concrete, the particular, the unique'. Pinnick and Gale (2000, p. 113) go one step further, maintaining that 'generalizations are not only unneeded, they are unwanted'.[11] Therefore, general philosophical analyses are not likely to advance history, since explaining particulars about the past is not philosophy's primary objective (Pinnick and Gale (2000, p. 110).

To sum up, critics think it unlikely that integrating History of Science and Philosophy of Science can succeed because, firstly, answering the questions addressed by historians of science and philosophers of science implies contrasting analyses, and, secondly, the lack of justification in the selection of historical cases and philosophical concepts clearly constitutes a methodological pitfall. They therefore conclude that it is impossible to produce work that contains good Philosophy of Science and good History of Science. In the rest of this chapter, I argue that this conclusion is premature and overly pessimistic. If we can manage to find research problems that fulfil the criteria introduced above, the matter of irrelevance disappears. The methodological challenges connected with the selection of historical cases and philosophical concepts are, however, real – just as real as they are in any study carried out by historians and philosophers. Fortunately, the selection of cases and concepts can be justified, while the relationship between the particular historical cases and the more abstract philosophical descriptions is less troublesome than it might first appear.

2.2 Understanding past research practice – a problem suited for iHPS

Explaining why scientists carried out their research in a particular way requires undertaking work that is both historiographical and philosophical in nature. This metascientific research problem is particularly suitable for iHPS as it entails two interdependent questions that lie at the heart of scholarship in History of Science and Philosophy of Science.

2.2.1 The historiographical question and its need for philosophy

What kind of research was undertaken in biological laboratories in the Dutch East Indies around 1900?[12] To what extent did the development of quantum theory depend on the earliest applications of quantum mechanics?[13] And how did researchers attempt to formulate, justify and criticise models of the photosynthetic reduction of carbon from 1840 to 1960?[14] These are all examples of legitimate historiographical questions, the answers to which contribute to our overall understanding of how scientists researched a particular topic at a given time and place.

But how can we explain the research activity of scientists? Gerd Graßhoff and Michael May (1995) argued that, in order to account for researchers' actions, one needs to know their goals, capacities (that is, their abilities and resources) and the norms that guided their actions. These factors together explain action A, given that the following principle holds: *If a researcher pursues goal G and believes, based on norm N, that action A is the best way to obtain G, and is capable of doing A, she will take action A.*[15] Research actions are geared towards solving the *research problem* at hand (see Figure 2.1). So, in other words, the way historical actors conceptualised the problem they wished to solve determined their actions.

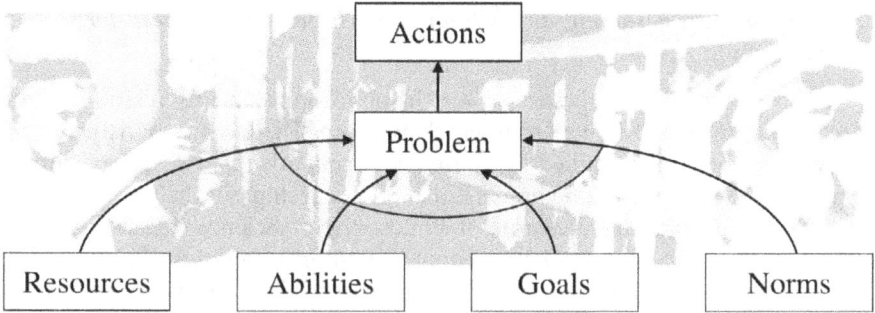

Figure 2.1 Schematic representation of how past research practice can be analysed and explained in terms of research problems. Research actions are directed towards solving the problem, whereas the problem is determined by the scientists' resources, abilities, goals and norms. BArch, Bild 183-K1004-0002 / Helmut Schaar / CC-BY-SA 3.0.

Researchers' goals typically resonate with their discipline's core concerns. In addition, we can expect that, from the wide range of principally interesting research questions, scientists choose those questions that they are capable of solving both in terms of infrastructure and technologies but also in terms of operational and conceptual know-how.[16] However, a research objective and the existing capacities together do not yet determine a research problem. The latter also defines what counts as an acceptable solution to the problem.[17] Hence, understanding past research practice requires that we know to which epistemic norms the practitioners conformed.

Some may wonder why philosophy is needed to work out the epistemic norms of the historical actors. Indeed, the claims made about researchers' goals, norms, and actions must first and foremost be based upon historical evidence. Nevertheless, philosophy can contribute greatly to the analysis of history: philosophical training helps scholars to uncover assumptions and expectations in the writing and actions of scientists (Pitt 2001, p. 379). Moreover, philosophical concepts can perform a heuristic function in historical research.[18] As Raphael Scholl and Tim Räz put it, 'philosophical concepts provide particularly pertinent questions that relate to science's core epistemic project'.[19] This applies especially to the objective of explaining past research practice, since the philosophical literature is filled with descriptions of abstract epistemic goals, norms and research strategies. Thirdly, I am inclined to argue for the integrative potential of philosophical concepts. Biophysicist Selig Hecht (1937, p. 227) wrote that 'one of the satisfactions in applying a theory comes from its unexpected illumination of data which, though well known, have long remained obscure in their interpretation and unrelated to the rest of the field'. In section 2.3, I argue that philosophical accounts can play an analogous role by suggesting that there are connections between certain historical phenomena that historians have so far overlooked.

2.2.2 The question pertaining to Philosophy of Science and its need for historiography

Let us now turn to the question asked in philosophy and its relation to the objective of understanding past research practice. Research actions are, as I have argued above, determined by epistemic norms. Identifying and explaining the norms that shape scientific practice, as well as carving out their premises and implications, are primary concerns of Philosophy of Science. In asking about norms, philosophers assume that researchers conform (perhaps unconsciously) to a set of standards relating to how to solve certain problems. A glance at history affirms this expectation. Researchers quite commonly specify the criteria that a satisfactory solution to a given problem should meet. Botanist Julius Sachs (1882), for example, wrote that 'in our present state of knowledge, formative processes in plants cannot be reduced to physico-chemical processes' and that the 'chain of causes cannot be determined in detail', thus leaving little room for doubt about the criteria that he believed satisfactory solutions to problems of plant morphogenesis should eventually fulfil.

There are two different approaches to identifying epistemic norms. We can either derive them from some first principles or we can extract them from actual scientific practice, which is exactly what history offers philosophy: access to past scientific practice.[20] For Nickles (1995, p. 147), 'history is a medium by which scientific precedents and test cases are made available to philosophy'. The disciplines of History of Science and Philosophy of Science are hence related in that the first provides data on research practice which can be used to assess philosophical claims about scientific inquiry.[21] In face of the criticism of this confrontational model, Scholl insists that the relationship between historical facts and philosophical theories should not be understood as a one-off confrontation. We do not frame general philosophical hypotheses which we test by their particular historical consequences. Nor do we generalise bottom-up from historical phenomena. Instead, iHPS studies move back and forth between what Hasok Chang has characterised as concrete (rather than particular) empirical exemplars and their abstract (rather than general) descriptions.[22] Through such a 'naturalised' approach, it is possible not only to identify philosophical norms, but also to trace how they have changed over the course of history.[23]

One could object, though, and claim that an *a priori* concept of science is needed to analyse past scientific research practice. However, there are alternatives. Nick Jardine (2000, p. 261) suggests making use of the fact that 'disciplines are regimes of organized behaviour inculcated and appropriated in educational institutions'. We can base our choice of historical actors on (contingent) historical criteria, such as actors' affiliation to institutions or networks of communication (for example, universities, societies, *respublica litteraria*, and so on), on the application of particular methods (for example, systematic observation or the method of difference) and also on researchers' contributions to particular research problems (such as planetary motion or heredity).

Another objection is that scientists are seldom philosophically consistent. As Nils Roll-Hansen (1992, p. 70) has pointed out, researchers 'may change their epistemological and methodological views over time, or even hold conflicting ideas at the same time. Or their philosophical tenets may conflict with their actual scientific beliefs or behaviour'. Nevertheless, Roll-Hansen insists that it is possible to analyse how researchers' methodological views guided their research practice.

In conclusion, we can state that as far as the objective of understanding past research practice is concerned, the questions asked in History of Science and Philosophy of Science are positively interdependent. On the one hand, a comprehensive historical account of past research practice requires that we take into consideration the methodological and epistemic norms to which the scientists conformed. On the other hand, the more we learn about scientists' conception of research problems, their research actions and capacities, the better we will be at discussing these norms. Explaining a historical actor's research activity surely requires extensive historical source work and detailed philosophical accounts (see section 4). But the effort involved in producing iHPS studies is of great value. The next section argues that iHPS work provides insights into issues untouched by mainstream History of Science and mainstream Philosophy of Science.

2.3 Research in physico-chemical biology between the two world wars

Both History of Science and Philosophy of Science deal with, among other things, the research practices of scientists. However, the two disciplines focus on different, barely overlapping aspects of this topic, and cross-references between the fields are few and far between.[24] As a result, historical and philosophical analyses tend to leave some interesting questions unanswered. To illustrate this point, this section reviews two separate debates held within the fields of History of Science and Philosophy of Science, roughly from the mid-1970s until the present time: the historiographical discussion about the so-called 'new biology' and the causes for its emergence during the first decades of the twentieth century, and the debate about mechanistic explanations, which gained momentum in Philosophy of Science around the turn of the twenty-first century. Integrating these analyses promises to advance our understanding of research in physico-chemical biology in the 1920s and 1930s.

2.3.1 Establishment of interwar biochemical and biophysical research

Between the two world wars, researchers, science managers and philanthropists on both sides of the Atlantic promoted the introduction of physical and chemical methods into the field of biology. Publication platforms, such as the *Journal of General Physiology* and *Monographs on Experimental Biology*, were founded for researchers interested in 'explaining life from the physico-chemical constitution of living matter' (Anonymous 1918, p. 217). Chairs and programmes for this kind of biology were created at several universities from the mid-1920s onwards, among them Cambridge, Columbia, Harvard, Princeton and Stanford,[25] while institutions devoted to research at the intersection of biology, chemistry and physics were established in, among other places, Dahlem, Pasadena, Paris and Copenhagen.[26]

Researchers working in these new laboratories used instruments and concepts developed in the physical sciences to address problems that traditionally had been worked on in physiology, as they believed that this would lead to progress in their respective fields.[27] Physiologist Josef Gicklhorn (1927, p. 92, my translation), for example, claimed that 'taking into consideration chemical as well as physical methods, laws and theorems in biology' was 'an absolute necessity and the most productive guiding principle for research'. Collaborative work between physicists, chemists and biologists became the rule at several newly established research institutions. On the occasion of the inauguration of the *Institut de Biologie Physico-Chimique* in Paris in 1930, biochemist Yvonne Khouvine (1930, p. 1051) explained that biology, physics and chemistry would be 'united in a single organization and thus be employed jointly in the study of problems which they could not solve separately'. A few years earlier, geneticist Thomas Hunt Morgan had similarly called for cross-disciplinary collaboration between biologists, chemists and physicists; in his view, biologists were needed in physical laboratories and physicists in biological laboratories.[28] Fritz Kögl (1930, p. 84), in his inaugural lecture as Professor of Organic Chemistry at Utrecht University, pointed out that 'to attack the difficult problems of the future, [organic chemists and biologists] must join forces'. Shortly afterwards, Kögl's group began collaborating with Friedrich Went, Utrecht's Professor of Botany, whose laboratory had long been studying plant movement and plant growth.[29]

2.3.2 Historians on the role of mechanistic thinking in interwar physico-chemical biology

Historians of science have shown an interest in physico-chemical biology and in the institutional policies that governed its implementation. Some have interpreted biologists' interest in collaborating with physicists and chemists as an attempt to establish biology as a proper science. For example, Garland Allen (2005, pp. 280–81) has described the mechanistic conception of biological phenomena as a strategy used by young biologists in the first decades of the twentieth century to distance themselves from old-fashioned natural history and its non-testable hypotheses. Others have emphasised the role of philanthropic institutions, particularly the Rockefeller Foundation and the director of its natural sciences division, Warren Weaver.[30] According to Robert Kohler, physicists and chemists were driven to address biological problems for financial reasons: researchers from the physical sciences collaborated with biologists to acquire funding for expensive instruments which otherwise they would not have been able to afford. In the resulting research collaborations, Kohler (1991, pp. 304–5) wrote, 'physical scientists provided technical services, and biochemists or biologists the problems'. More recently, Michel Morange (2007, p. 108) has pointed to the larger political dimension of the philanthropic support of the biological sciences, referring to a widely shared notion in the 1930s that biology should provide solutions for social problems.

Some scholars have also made references to certain research strategies and scientists' methodological norms. Evelyn Fox Keller, for example, has concluded that the introduction of physics to the field of biology in the 1930s led to a transformation of language, focus and methodology. According to Fox Keller (1990, pp. 406–7), this transformation resulted in an entire new set of beliefs – among them the belief in 'a particular kind of (linear, causal) mechanism'. Mechanisms also play a role in Daniel Kevles and Gerald Geison's (1995, p. 101) account of the experimental life sciences. In particular, they have ascribed the privileged status of fields such as biochemistry and neurophysiology in the 1930s to these disciplines' ability to demonstrate the 'fertility of the mechanistic approach to biological problems – specifically, by showing the extent to which extremely complex events ... could be explained by or "reduced to" electrical-chemical and other basic physical concepts'. Soraya de Chadarevian and Harmke Kamminga (1998, p. 5) have emphasised that researchers who 'reduced bodily functions to the interplay of molecules' made efforts to develop technologies 'to measure and monitor molecules in the body'. However, usually historians did not specify in detail what the belief in causal mechanism implied or why exactly it was believed that in fields like biochemistry biological phenomena were, in fact, successfully reduced to physico-chemical concepts. Some philosophers of science, by contrast, have endeavoured to clarify these issues.

2.3.3 Philosophers on mechanisms and scientific practice

More than forty years ago the philosopher William C. Wimsatt (1976, p. 671) claimed that most biologists 'see their work as explaining types of phenomena by discovering mechanisms', and that this is referred to as reduction. William Bechtel and Andrew Hamilton (2007) have, more recently, specified the way in which mechanistic explanations are reductive. They characterise mechanisms as entities and activities organised in such a way that they are responsible for the phenomenon to be explained.[31] Now, putting forward mechanistic explanations is reductive in that it involves 'decomposing the system responsible for a phenomenon into component parts and component operations' (Bechtel and Hamilton 2007, p. 405). The new mechanical philosophy thus focuses on an abstract research goal: the objective of discovering how the phenomenon of interest is produced. By specifying what constitutes an adequate description of a mechanism, philosophers outlined a number of general epistemic norms.[32]

Assertions about scientific practice were able to be made from this characterisation of mechanisms – inferences that were justified by the assumption that 'in the discovery of mechanisms, the product shapes the process of discovery' (Darden 2017, p. 264). Philosophers inferred, for instance, that, in order to put forward an explanation, researchers have to identify the entities and activities of the mechanism, assign entities to activities and show how these components are organised, so that, under specific contextual conditions, the mechanism produces the phenomenon to be explained (Bechtel 2009, p. 553). Philosophers also identified a series of reasoning and experimental strategies to guide the discovery of mechanisms.[33] These strategies are abstract descriptions of research actions which help scientists find the mechanism responsible for the phenomenon of interest – that is, they help scientists achieve the goal of explaining a biological phenomenon in an acceptable way. Last but not least, mechanistic research is claimed to have an integrative effect. According to philosophers Carl Craver and Lindley Darden (2013, pp. 162–3), the 'integration of biology is forged by building mechanism schemas that … satisfy evidential constraints from many areas of biology (chemistry and physics too)'. Since different fields are suited to the study of different kinds of evidential constraints, the search for mechanisms promotes interfield integration and collaborations across disciplinary boundaries.

In conclusion, the debate on mechanisms held over the past twenty years in Philosophy of Science covers many of the topics raised in historians' analyses of interwar physico-chemical biology. However, the significance of the new mechanical philosophy is based on the assumption that the search for mechanisms was (and still is) important in actual scientific practice. This assumption cannot be supported by philosophical considerations alone. Rather, historical evidence – for instance, from biophysical and biochemical research which was carried out in the 1920s and 1930s – is better suited to substantiating such a hypothesis.

2.3.4 Unanswered questions and the promise of integration

Historians and philosophers of science have engaged extensively with physico-chemical biology and mechanistic explanations. Nevertheless, some interesting questions remain unanswered. Firstly, it is not precisely known where, when and how researchers came to pursue this type of research. Craver and Darden (2013, p. 3), for instance, were unable to uncover in their recent research 'just when and how this mechanistic view of science entered the different fields of biology, specifically, and precisely how the idea of mechanism came to so thoroughly triumph as a way of thinking about explanation in biology'. Secondly, the philosophical literature provides more of a sketch of an account of scientific practice than a comprehensive analysis. There is still much to learn about the conceptual, material and social preconditions and implications of mechanistic research. Historical analyses will allow scholars to assess the impact of the epistemic norms worked out in the new mechanical philosophy, since by studying history we can learn more about the prevalence and importance of mechanistic reasoning in science. Similarly, by analysing concrete instantiations of past cross-disciplinary research, we can assess the conclusiveness of philosophical statements about interfield integration.

Similarly, most historians' analyses tell us little about the theoretical and methodological assumptions of cross-disciplinary research in physico-chemical biology between the world wars. Many accounts affirm the importance of Claude Bernard's or Jacques Loeb's methodological views, and show how these views inspired leading scientists and philanthropists to promote a certain style of research.[34] There, however, the accounts end. For example, it is not yet agreed how scientists came to realise that certain methods could help researchers find solutions to specific biological problems. It is also not entirely clear why biologists and practitioners from the physical sciences decided to collaborate in the first place and how this affected their experimental procedures and theoretical models. The need to know the answers to these questions would be less urgent if it was clear that the emergence of cross-disciplinary research always hinged upon special infra-structures and funding opportunities.[35] However, this is clearly not the case. Went and Kögl's joint research project, for example, was not funded by the Rockefeller Foundation, nor were there institutional incentives for cross-disciplinary research at Utrecht University. Such cases rather suggest that researchers sometimes had other reasons – possibly epistemic ones – for collaborating with practitioners from other fields.

Although epistemic factors seldom lie at the heart of historians' analyses, there is no methodological reason for this bias. After all, recognising the significance of non-epistemic factors does not render the study of epistemic factors obsolete. Whatever it was that motivated researchers to enter the field of physico-chemical biology, eventually they had to decide which problems to solve and how; it was at this point that epistemic norms came into play. We have good reason to believe that in planning their research scientists took into account the criteria that would need to be met for their peers to accept their new solutions. The idea of binding disciplinary norms is not unknown to historians. Quite the contrary: a consensus of what constitutes an interesting question, an appropriate investigation, or an acceptable answer is an integral part of most definitions of the term 'discipline'.[36] Historians of science have found that young scientists are routinely 'instructed' during their training to follow a certain investigative protocol: they learn how to achieve and present their findings in a way that makes their results acceptable to their peers. Whether the research in physico-chemical biology in the early decades of the twentieth century was really shaped by norms similar to those discussed in the new mechanical philosophy is, of course, an empirical question.

In summary, historians and philosophers examined similar points: (a) the search for explanations in which biological phenomena are 'reduced' to physico-chemical events and the properties of living matter; (b) the use of physico-chemical methods in biology; and (c) cross-disciplinary research collaborations. While historians presented these as separate features of research in physico-chemical biology, philosophers offered a unifying explanation: advocates of the new mechanistic philosophy see features (a)-(c) as the consequences of the goal of scientists to find mechanisms. It might eventually be possible to explain researchers' experiments and their hypotheses as well as their decisions to collaborate with researchers from other fields as a direct result of their desire to discover mechanisms. Hence, philosophy has indeed the potential to advance our understanding of past research practice. In the following and final section, I discuss how to avoid the methodological pitfalls discussed in section 2, how to identify historical actors' research problems, goals, capacities, norms and actions. I then conclude this chapter by pointing out the relevance of such studies for History of Science and Philosophy of Science.

2.4 How to learn about past research practice through iHPS

Before addressing a metascientific research problem by integrating historical and philosophical analyses, it is necessary to establish that the question being dealt with does, in fact, constitute a suitable research topic for iHPS. That is, the research problem must draw attention to both philosophical and historiographical questions. We saw that opponents of iHPS believe that the very selection of philosophical issues and historical events is problematic and leads to methodological flaws. I will counter this criticism by showing that such selections can be justified.

2.4.1 Selecting historical instances and philosophical frameworks

In the previous section, I argued that the new mechanical philosophy offers a promising explanatory framework to scholars interested in uncovering the motivation behind the decision of researchers to combine physico-chemical and biological methods. Of course, we should be careful to avoid historical anachronism, a historically incoherent interpretation of the writings and deeds – and we might add, the actions – of the historical figures. According to Jardine (2000, p. 252), being aware of the 'material, psychological, social and institutional conditions of the production of deeds and works' helps us avoid vicious anachronism. If nothing else, as Jardine explains, the presuppositions of our modern analytical categories – for example the modern category mechanism as characterised in the new mechanical philosophy – should be realised by the group of actors under consideration. Jardine (2000, pp. 254–5) demands that 'the categories we apply to past deeds and works should not be at odds with the entire range of the significances attached to them in the past societies in question'. As to our example, we need to establish that the presuppositions of the new mechanical philosophy hold good for the researchers under consideration (ibid., p. 266). We should be able to show, for instance, that botanist Went did indeed believe that in order to explain plant growth he needed to describe the relevant entities involved in plant growth, work out their physico-chemical properties and activities, and figure out how they were organised in such a way that the phenomenon was produced.

The influence of epistemic norms on research practice can best be assessed in cases where the research has been planned from scratch, since the influence of contingent factors will then be minimal. Setting up a new research laboratory comes remarkably close to a state of *tabula rasa*. When Thomas H. Morgan agreed to organise Caltech's biology department in 1927, for example, his choice of goals, resources and abilities was relatively unconstrained.[37] He was free to choose which fields (and therefore which goals and norms) to include, which instruments to buy (and thereby what resources to provide) as well as who to employ (that is, he could decide the kind of abilities that were needed), and he would have made these choices in accordance with his normative views on what constitutes good research. Moreover, we can be pretty certain that Morgan made his decisions cautiously, knowing that his choices would determine research at Caltech for the next few years at least.

Morgan (1927, p. 86) was eager to establish research projects at the intersection of the biological and physical sciences. He offered the post of Assistant Professor of General Physiology to a young plant physiologist from the Netherlands, Herman E. Dolk. A former assistant to Went, Dolk knew how to perform an intricate plant test that was crucial for isolating growth hormones.[38] Another European scientist engaged by Morgan was the English biochemist Kenneth V. Thimann. After their arrival in Pasadena in the spring of 1930, Thimann and Dolk worked together on plant growth substances until Dolk's untimely death two years later.[39] Analysing the early phase of such a cross-disciplinary research project can be illuminating. Collaborative work can generate historical sources that are especially revealing about the researchers' goals, norms, and capacities: in the above case, for example, the scientists needed to clarify and align their theoretical assumptions, methodological norms, and expertise in order to ensure that joining forces would be effective; and they had to explain their new concepts and methods of working to scientists working in their own discipline.

In sum, philosophical frameworks and historical actors are not selected arbitrarily: we deliberately choose philosophical frameworks with high explanatory power regarding the metascientific research problem to be solved; these frameworks' general assumptions need to be in accordance with what we know about the goals, beliefs and actions of the scientists under consideration; while we focus on historical actors for whom the phenomena we wish to examine can be expected to be well documented.

2.4.2 Identifying scientists' research problems, actions, goals, capacities and norms

Once a selection of research projects has been made, the work practices of the scientists and the factors that determined their research problems can be uncovered. The historical sources that allow us to reconstruct scientists' resources include publications, correspondence, laboratory inventories, receipts, instruments, photographs and films. Morgan, for instance, wrote to Went describing the small laboratory with two underground rooms – 'suitable for carrying out ... work on growth substances' – that he had arranged, while Dolk's correspondence with Went even included floorplans of these facilities.[40] Occasionally, researchers discussed their abilities explicitly. In his inaugural lecture of 1930, Kögl explained that, unlike chemists, biologists are able to develop physiological tests. Organic chemists, on the other hand, have an idea of what needed to be done to purify hormones.[41] In other cases, one can derive the importance of abilities indirectly: after Dolk died in a car accident in 1932, Morgan declared that they would not be able to continue their phytohormone research 'unless someone as competent as Doctor Dolk could be found to take his place'.[42]

To map a field's goals at any given time, we can consult contemporary textbooks, programmatic lectures and conference programmes. More specific goals can be identified in publications, especially in reviews. In a review of 1927, botanist Peter Stark (1927, p. 91) referred to the isolation of growth substances as a task 'whose fulfilment ranks among the most urgent research desiderata'. Moreover, researchers' conceptualisation of problems can, for example, be gleaned from their grant proposals and publications. In their first joint article, Dolk and Thimann (1932, pp. 30–31) wrote that in order to get a better idea of the mechanism of the growth substance's action, it was necessary to find out more about its chemical structure.

Fortunately, norms can be taken from facts, and a discipline's abstract norms can be extracted from method discussions, especially those in textbooks.[43] Plant physiologist Wilhelm Pfeffer (1900, p. 4), for example, stated:

> In explaining the relations between cause and effect, Physiology ... must determine the nature and properties of the different parts affected, as well as the accompanying external factors which may be involved. Both the character of the stimulus to action and the necessary mechanical means by which the resulting phenomenon is affected require to be known ....

Went (1933, p. 137), for his part, wrote: 'Investigating a factor's impact on a life process seems to be comparatively simple. All that has to be done is to keep all other factors constant, vary the one, and see the result.' Passages like these reveal scientists' views on the criteria that needed to be met to support a claim with adequate evidence. As Jutta Schickore (2017, p. 5) puts it, 'methods-related concepts, statements, and reflections as they are presented in experimental reports are significant because they reflect the authors' understanding of the structure and organization of good experimental research'. Certainly, we cannot take scientists' programmatic methodological pronouncements as a reliable account of the norms that shaped their actions. To identify the latter, we need to focus on the scientists' actions and subsequently try to establish that these were governed by specific norms. This kind of procedure is not uncommon in historiography.[44] In turn, researchers' actions can be worked out with the help of sources such as laboratory notebooks, correspondence, internal reports and publications. The latter can contain helpful clues concerning researchers' intentions. Take, as an example, one of Went's students, who described his experimental work and the reasoning that provided him with 'data concerning the mechanism of the action of the [plant growth] hormone' as follows:

> By applying the substance which promotes elongation *and simultaneously preventing the cell from elongation*, it proved to be possible to concentrate on the first phases of the process of cell elongation .... Hence it was possible to distinguish between causes and effects of elongation, which had always been the greatest difficulty in the analysis of growth.
>
> (Heyn 1933, pp. 78–79)

2.4.3 Assessment of the findings' relevance to History of Science and Philosophy of Science

It is hard to see how investigating scientists' research actions and their conceptualisation of research problems cannot be relevant to the History of Science. Such investigations provide insights into the material, institutional, operational and social conditions of past scientific practice, while also emphasising the role of certain abilities and the significance of new technologies in answering long-standing research questions. It is also interesting – particularly for philosophers – to learn how robust the epistemic norms under consideration have been over time. This can be evaluated by comparing the norms that shaped phytohormone research with the norms discussed in earlier scientific work on similar problems. The fact that Went's work was so well received in turn substantiates the claim that the standards concerned were widely accepted at the time. We can thus learn about the prevalence of the norms exposed by the new mechanistic philosophy in actual scientific practice.

Crucially, iHPS studies do not end with the investigations of particular local cases of past scientific work. As Scholl (2018, p. 237) puts it, we strive 'to know how broadly, and in what variations, certain abstract descriptions apply to concrete instances'. Therefore, we should try to obtain more abstract descriptions from our concrete historical studies, since by studying additional concrete historical cases we will be in a better position to assess the empirical scope of these abstract descriptions.[45]

Conclusion: The case for iHPS

I have argued that integrating philosophical and historical analyses is suitable for addressing metascientific research problems which comprise questions pertaining to History of Science that cannot be answered without turning to philosophy as well as philosophical concerns that require detailed historical analysis. Understanding past research practice is an ideal research problem for iHPS: it includes the goal of understanding why practitioners of a certain field working at a certain time and place conducted their research on a certain topic in a particular way, as well as the goal of identifying and explaining effective methodological norms. Properly pursued, iHPS scholarship advances both our philosophical and our historical understanding. The more we attend to the material and social conditions of past research practice, the easier it will be to bring to light the epistemic norms of historical actors. And the more we know about these norms, the easier it will be for us to understand the actions that scientists undertook to solve their research problems.

I will close this chapter by quoting biophysicist Selig Hecht again: 'We have the foolish, but understandable, philosophical notion that there are some conditions under which natural phenomena can be investigated. And the problem [is] to provide the conditions.'[46] As historians and philosophers of science, what is so wrong with delving into the conditions that researchers considered critical for their investigations? Doing so does not imply that identifying the epistemic norms of past scientific actions is the only or even primary purpose of engaging in the history of science. But they should not be neglected either, certainly not for methodological reasons. There is still much to learn about the role of methodological principles in the development of science; if this challenge were taken up, it could be termed a 'third area', a 'science of methodology', 'scientonomy', or simply 'good History of Science and good Philosophy of Science'.[47]

Notes

1 Anonymous, *Memorandum*, undated (probably mid-1940s), Cambridge University Library, BCHEM 4, folder 12. For the early development of History of Science at Cambridge University, see Anna-K. Mayer (2000) and (2004). The viewpoint of Cambridge University's Faculty Board was possibly influenced by the programme of William Whewell, Ernst Mach and Pierre Duhem. These early students of science all turned to history to evaluate philosophical claims about science. This has been pointed out by, among others: Larry Laudan and Rachel Laudan (2016); Thomas Nickles (1995, p. 143); and I. Bernard Cohen (1974).
2 On the first statement, see Danto (1962). On the second statement, see, e.g., Joseph Pitt (2001).
3 Committee for Integrated HPS, *Mission statement on integrated HPS*, http://inte gratedhps.org/en/about/ (accessed 3 April 2018).
4 See, e.g. Hanne Andersen and Susann Wagenknecht (2013); Lindley Darden and Nancy Maull (1977); Jane Maienschein (1993); K. Brad Wray (2002).
5 E.g. Pitt (2001), pp. 373–82; Cassandra Pinnick and George Gale (2000); Ronald N. Giere (1973).
6 For more comprehensive discussions, see the contributions in *The Philosophy of Historical Case Studies*, ed. by Tilman Sauer and Raphael Scholl (2016), and *Integrating History and Philosophy of Science: Problems and Prospects*, ed. by Seymour Mauskopf and Tad Schmaltz (2012).
7 For a more comprehensive discussion of this point, see Raphael Scholl (2018), pp. 228–232.
8 On this point, see, e.g. L. Pearce Williams (1975); Larry Laudan et al., (1986) p. 158.
9 Hanson (1962, p. 581) wrote that history only describes the actions of the historical actors, but it cannot evaluate the actions or views of the historical actors.
10 Currie and Walsh's forthcoming paper revolves around this issue.
11 Cf. Ulrich Charpa (1985), who argues that the focus on dichotomies is misleading.
12 See Robert-Jan Wille, (2015).

13 See Christian Joas and Jeremiah James (2015).
14 See Kärin Nickelsen (2015).
15 See Gerd Graßhoff and Michael May (1995, p. 58), an account that was inspired by Paul Churchland (1970).
16 On how certain techniques transformed how biologists conceptualised problems and organised their work, see Angela N. H. Creager and Hannah Landecker (2009).
17 See Thomas Nickles (1981, p. 113); Alan C. Love (2008, p. 875).
18 See, e.g., Jutta Schickore (2011).
19 See Raphael Scholl and Tim Räz (2016, p. 88).
20 Ernan McMullin (1976, pp. 593, 599) emphasised this point, referring to the history of science as 'a body of information about the past of science' and that 'there is no alternative to invoking the historical record'; Laudan and Laudan (2016, p. 77) stress that 'without empirical checks, it is hard to see how generalisations about the evolution of science can be more than just-so stories'.
21 This 'confrontational model' is discussed in this volume by Matteo Vagelli in Chapter 5, and by Klodian Coko in Chapter 10.
22 Scholl, (2018). That the relationship between historical and philosophical studies should be seen as one between the concrete and the abstract, rather than between the particular and the general was first raised by Hasok Chang (2011).
23 On 'naturalised' Philosophy of Science, see Greg Rupik and Matteo Vagelli's Chapters 1 and 5 in this volume; see also: William Bechtel (2009, p. 549); Bechtel and Richardson (2010); Lindley Darden (1991), pp. 15–17. According to historian of biology Marga Vicedo (1992, p. 495), 'our epistemic values and our understanding of the world are the results of the contingent facts of history'. Cohen (1974, p. 312) agrees, believing 'that a greater sensitivity to the cannons of history could readily produce a dramatic rise in the historical level of accuracy and authenticity of philosophical discourse'.
24 Among the few works investigating historical actors' research strategies is Kärin Nickelsen's book on photosynthesis, cited above, and Raphael Scholl and Nickelsen (2015).
25 See Robert E. Kohler 1982, p. 308.
26 See, e.g.: Finn Aaserud (1990); Michel Morange (2007); Doris T. Zallen (1992).
27 This group comprised researchers working under various disciplinary labels, such as physico-chemical biology, general physiology, experimental biology, biochemistry, biophysics, physiological chemistry, etc.
28 See Thomas H. Morgan (1927, pp. 214, 217).
29 See Caterina Schürch (2017); and Patricia Faasse(1994).
30 Pnina Abir-Am (1982); Robert E. Kohler (1991); Lily E. Kay (1993).
31 Bechtel and Hamilton (2007, p. 405). A concise account of the different characterisations of mechanisms discussed in the new mechanical philosophy is provided by Carl Craver and James Tabery (2017).
32 For detailed account of these norms, see Chapter 4 of Carl F. Craver (2007), especially pp. 123–52.
33 Research strategies are discussed, e.g., in: William Bechtel and Robert C. Richardson (2010); Carl F. Craver (2002); Craver and Bechtel (2007; Lindley Darden (2002); Darden and Craver (2002).
34 E.g. Philip J. Pauly (1987).
35 On the role of science policy in scientific research and its promise for iHPS, see Eugenio Petrovic's contribution in Chapter 4 of this volume.
36 See, e.g., Kathryn M. Olesko (1991), p. 14; Kohler (1982), pp. 1–2; William Bechtel (2005), pp. 7–8; Geoffrey E. R. Lloyd (2009), p. 170, and Alex Aylward's contribution in Chapter 11 of this volume.
37 See, e.g., Kay 1993, Chapter 3.

38　For more on the *Avena* test, see Schürch 2017, pp. 8–16.
39　See Herman E. Dolk and Kenneth V. Thimann (1932).
40　Morgan to Went, 25 March 1932. Leiden, Boerhaave Museum, Correspondence of F.A.F.C. Went, a79, file 'Morgan'; and Dolk to Went, 20 February 1931. Leiden, Boerhaave Museum, Correspondence of F.A.F.C. Went, a79, file 'Dolk, H. E.'.
41　Fritz Kögl 1930, pp. 82–85.
42　Morgan to Went, footnote 40.
43　On this point, see also Nickles 1995, p. 146, who underlines that 'the consequent of the conditional makes a factual claim that can be checked empirically'. In this comment, Nickles refers to Laudan's claim that methodological rules can be reformulated as conditionals of the form, 'If you want to achieve goal G, then procedure P is a reliable way to do so'.
44　See, e.g., Barbara Duden (1987).
45　Ibid., p. 226. See also Claudia Cristalli's discussion of narratives as frameworks that have the potential of bridging the gap between accounts of particular historical events and bigger, more abstract processes in the next chapter.
46　Hecht to Crozier, 22 June 1926. Harvard, Pusey Library, Papers of William John Crozier, HUG 4308.5, Box 3, file 'Hecht S. 1923–27'.
47　The first two terms listed are from Giere 1973, p. 294. On scientonomy, see Greg Rupik's contribution in Chapter 1 of this volume.

Acknowledgements

I thank the editors for their encouragement and very helpful feedback on the chapter. I am grateful to Raphael Scholl, David Munns as well as the participants of the 2017 Cambridge Graduate Seminar 'Aims and Methods of Histories of the Sciences' for comments on an earlier version of the chapter. Special thanks go to Kärin Nickelsen for supervising and supporting this work and to Margareta Simons for editing the manuscript. This work was partly supported by the Swiss National Science Foundation (grant number P1SKP1_174773).

Bibliography

Aaserud, Finn, *Redirecting Science: Niels Bohr, Philanthropy, and the Rise of Nuclear Physics* (Cambridge: Cambridge University Press, 1990).

Abir-Am, Pnina, 'The Discourse of Physical Power and Biological Knowledge in the 1930s: A Reappraisal of the Rockefeller Foundation's "Policy" in Molecular Biology', *Social Studies of Science*, 12. 3(1982), 341–382.

Allen, Garland E., 'Mechanism, Vitalism and Organicism in Late Nineteenth- and Twentieth-century Biology: The Importance of Historical Context', *Studies in History and Philosophy of Biological and Biomedical Sciences*, 36. 2(2005), 261–283.

Andersen, Hanne and Susann Wagenknecht, 'Epistemic Dependence in Interdisciplinary Groups', *Synthese*, 190. 11(2013), 1881–1898.

Anonymous, 'Publications on Experimental Biology and General Physiology', *Science*, 48. 1235 (1918), 217.

Bechtel, William, *Discovering Cell Mechanisms: The Creation of Modern Cell Biology* (Cambridge: Cambridge University Press, 2005).

Bechtel, William, 'Constructing a Philosophy of Science of Cognitive Science', *Topics in Cognitive Science*, 1. 3(2009), 548–569.

Bechtel, William and Andrew Hamilton, 'Reduction, Integration, and the Unity of Science: Natural, Behavioral, and Social Sciences and the Humanities', in *General Philosophy of Science: Focal Issues*, ed. by Theo Kuipers (Amsterdam: Elsevier, 2007), 377–430.

Bechtel, William and Robert C. Richardson, *Discovering Complexity: Decomposition and Localization as Strategies in Scientific Research*, 2nd edition (Cambridge, MA/London: MIT Press, 2010).

Chang, Hasok, 'Beyond Case-Studies: History as Philosophy', in *Integrating History and Philosophy of Science*, ed. by Seymour Mauskopf and Tad Schmaltz (Dordrecht: Springer, 2011), 109–124.

Charpa, Ulrich, 'On Drawing Distinctions between History and Philosophy of Science', *Erkenntnis*, 23. 3(1985), 251–253.

Churchland, Paul, 'The Logical Character of Action-Explanations', *The Philosophical Review*, 79. 2(1970), 214–236.

Cohen, I. Bernard, 'History and the Philosopher of Science', in *The Structure of Scientific Theories*, ed. by Frederick Suppe (Urbana/Chicago: University of Illinois Press, 1974), 308–349.

Committee for Integrated HPS, *Mission Statement on Integrated HPS*, http://integratedhps.org/en/about/ (accessed 3 April 2018).

Craver, Carl F., 'Interlevel Experiments and Multilevel Mechanisms in the Neuroscience of Memory', *Philosophy of Science*, 69. S3(2002), S83–S97.

Craver, Carl F., *Explaining the Brain: Mechanisms and the Mosaic Unity of Neuroscience* (Oxford: Oxford University Press, 2007).

Craver, Carl F. and William Bechtel, 'Top-down Causation without Top-down Causes', *Biology and Philosophy*, 22. 4(2007), 547–563.

Craver, Carl F. and Lindley Darden, *In Search of Mechanisms: Discoveries across the Life Sciences* (Chicago/London: University of Chicago Press, 2013).

Craver, Carl F. and James Tabery, 'Mechanisms in Science', in *The Stanford Encyclopedia of Philosophy*, ed. by Edward N. Zalta, Spring 2017 edition (Stanford: Metaphysics Research Lab, 2017), URL: https://plato.stanford.edu/archives/spr2017/entries/science-mechanisms/

Creager, Angela N. H. and Hannah Landecker, 'Technical Matters: Method, Knowledge and Infrastructure in Twentieth-Century Life Science', *Nature Methods*, 6. 10 (2009), 701–705.

Currie, Adrian and Kirsten Walsh, 'Frameworks for Historians & Philosophers', *HOPOS*, (Forthcoming).

Danto, Arthur C., 'Narrative Sentences', *History and Theory* 2. 2(1962), 146–179.

Darden, Lindley, *Theory Change in Science: Strategies from Mendelian Genetics* (New York: Oxford University Press, 1991).

Darden, Lindley, 'Strategies for Discovering Mechanisms: Schema Instantiation, Modular Subassembly, Forward/Backward Chaining', *Philosophy of Science*, 69. S3 (2002) S354–365.

Darden, Lindley, 'Strategies for Discovering Mechanisms', in *The Routledge Handbook of Mechanisms and Mechanical Philosophy*, ed. by Stuart Glennan and Phyllis Illari (London/New York: Routledge, 2017), 255–266.

Darden, Lindley and Carl Craver, 'Strategies in the Interfield Discovery of the Mechanism of Protein Synthesis', *Studies in History and Philosophy of Science Part C: Studies in History and Philosophy of Biological and Biomedical Sciences*, 33. 1 (2002), 1–28.

Darden, Lindley and Nancy Maull, 'Interfield Theories', *Philosophy of Science*, 44. 1(1977), 43–64.

de Chadarevian, Soraya and Harmke Kamminga, 'Introduction', in *Molecularizing Biology and Medicine: New Practices and Alliances, 1910s–1970s*, ed. by ibid. (Amsterdam: Harwood Academic Publishers, 1998), 1–16.

Deutsch, Morton, 'Cooperation and Competition', in *The Handbook of Conflict Resolution: Theory and Practice*, ed. by Morton Deutsch, Peter T. Coleman, and Eric C. Marcus, 2nd edn (San Francisco: Jossey-Bass, 2006), 23–42.

Dolk, Herman E. and Kenneth V. Thimann, 'Studies on the Growth Hormone of Plants: I', *Proceedings of the National Academy of Sciences of the United States of America*, 18(1932), 30–46.

Duden, Barbara, *Geschichte unter der Haut. Ein Eisenacher Arzt und seine Patientinnen um 1730* (Stuttgart: Klett-Cotta, 1987).

Faasse, Patricia, *Experiments in Growth* (Amsterdam: De Jong, 1994).

Fox Keller, Evelyn, 'Physics and the Emergence of Molecular Biology: A History of Cognitive and Political Synergy', *Journal of the History of Biology*, 23. 3(1990), 389–409.

Gicklhorn, Josef, 'Mikrochemie und Mikrophysik: Ihre biologische Auswertung in Gegenwart und Zukunft', *Protoplasma*, 2. 1(1927), 89–125.

Giere, Ronald N., 'History and Philosophy of Science: Intimate Relationship or Marriage of Convenience?', *The British Journal for the Philosophy of Science*, 24. 3 (1973), 282–297.

Graßhoff, Gerd and Michael May, 'Methodische Analyse wissenschaftlichen Entdeckens', *Kognitionswissenschaft*, 5(1995), 51–67.

Hanson, Norwood Russell, 'The Irrelevance of History of Science to Philosophy of Science', *The Journal of Philosophy*, 59. 21(1962), 574–586.

Hecht, Selig, 'The Instantaneous Visual Threshold after Light Adaptation', *Proceedings of the National Academy of Sciences of the United States of America*, 23. 4(1937), 227–233.

Heyn, Antonius N. J., 'Further Investigations on the Mechanism of Cell Elongation and the Properties of the Cell Wall in Connection with Elongation I: The Load Extension Relationship', *Protoplasma*, 19(1933), 78–97.

Jardine, Nick, 'Uses and Abuses of Anachronism in the History of the Sciences', *History of Science*, 38. 3(2000), 251–270.

Joas, Christian and Jeremiah James, 'Subsequent and Subsidiary? Rethinking the Role of Applications in Establishing Quantum Mechanics', *Historical Studies in the Natural Sciences*, 45. 5(2015), 641–702.

Kay, Lily E., *The Molecular Vision of Life: Caltech, the Rockefeller Foundation, and the Rise of the New Biology* (New York: Oxford University Press, 1993).

Kevles, Daniel J. and Gerald L. Geison, 'The Experimental Life Sciences in the Twentieth Century', *Osiris*, 10(1995), 97–121.

Khouvine, Yvonne, 'The New French Institute of Physico-Chemical Biology', *Journal of Chemical Education*, 7. 5(1930), 1051–1057.

Kögl, Fritz, 'Wege und Ziele der Erforschung von Naturstoffen. Antrittsrede', in *Jaarboek der Rijks-Universiteit te Utrecht 1930–1931* (Utrecht: J. van Druten, 1930), 73–91.

Kohler, Robert E., *From Medical Chemistry to Biochemistry: The Making of a Biomedical Discipline* (Cambridge: Cambridge University Press, 1982).

Kohler, Robert E., *Partners in Science. Foundations and Natural Scientists 1900–1945* (Chicago/London: University of Chicago Press 1991).

Kuukkanen, Jouni-Matti, 'Historicism and the failure of HPS', *Studies in History and Philosophy of Science Part A*, 55(2016), 3–11.

Laudan, Larry and Rachel Laudan, 'Scientific Change: Philosophical Models and Historical Research', *Synthese* 69. 2(1986), 141–223.

Laudan, Larry and Rachel Laudan, 'The Re-emergence of Hyphenated History-and-philosophy-of-science and the Testing of Theories of Scientific Change', *Studies in History and Philosophy of Science Part A*, 59(2016), 74–77.

Lloyd, Geoffrey E. R., *Disciplines in the Making: Cross-Cultural Perspectives on Elites, Learning, and Innovation* (Oxford: Oxford University Press, 2009).

Love, Alan C., 'Explaining Evolutionary Innovations and Novelties: Criteria of Explanatory Adequacy and Epistemological Prerequisites', *Philosophy of Science*, 75. 5(2008), 874–886.

Maienschein, Jane, 'Why Collaborate?', *Journal of the History of Biology*, 26. 2(1993), 167–183.

Mauskopf, Seymour and Tad Schmaltz, eds, *Integrating History and Philosophy of Science: Problems and Prospects* (Dordrecht: Springer, 2012).

Mayer, Anna-K., 'Setting up a Discipline: Conflicting Agendas of the Cambridge History of Science Committee, 1936–1950', *Studies in History and Philosophy of Science Part A*, 31. 4(2000), 665–689,

Mayer, Anna-K., 'Setting up a discipline, II: British history of science and "the end of ideology", 1931–1948', *Studies in History and Philosophy of Science Part A*, 35. 1 (2004), 41–72.

McMullin, Ernan, 'History and Philosophy of Science: A Marriage of Convenience?', *PSA 1974: Proceedings of the 1974 Biennial Meeting Philosophy of Science Association* (1976), 585–601.

Morange, Michel, 'Physics, Biology and History', *Interdisciplinary Science Reviews*, 32. 2(2007), 107–112.

Morgan, Thomas H., 'The Relation of Biology to Physics', *Science*, 65. 1679(1927), 213–220.

Morgan, Thomas H., 'Study and Research in Biology', *Bulletin of the California Institute of Technology: Annual Catalogue*, 36. 117(1927), 86–91.

Nickelsen, Kärin, *Explaining Photosynthesis: Models of Biochemical Mechanisms, 1840–1960* (Dordrecht: Springer, 2015).

Nickles, Thomas, 'What Is a Problem that We May Solve It?', *Synthese*, 47. 1(1981), 85–118.

Nickles, Thomas, 'Philosophy of Science and History of Science', *Osiris*, 10(1995), 138–163.

Olesko, Kathryn M., *Physics as a Calling: Discipline and Practice in the Koenigsberg Seminar for Physics* (Ithaca/London: Cornell University Press, 1991).

Pauly, Philip J., *Controlling Life: Jacques Loeb and the Engineering Ideal in Biology* (New York/Oxford:Oxford University Press, 1987).

Pfeffer, Wilhelm, *The Physiology of Plants: A Treatise upon the Metabolism and Sources of Energy in Plants*, Vol. 1, 2nd edn (Oxford: Clarendon Press, 1900).

Pinnick, Cassandra and George Gale, 'Philosophy of Science and History of Science: A Troubling Interaction', *Journal for General Philosophy of Science*, 31. 1(2000), 109–125.

Pitt, Joseph, 'The Dilemma of Case Studies: Toward a Heraclitian Philosophy of Science', *Perspectives on Science*, 9. 4(2001), 373–382.

Roberts, Clayton, *The Logic of Historical Explanation* (University Park PA: Pennsylvania State University Press, 1996).

Roll-Hansen, Nils, 'Philosophical Ideas and Scientific Practice: A Note on the Empiricism of T.H. Morgan', *Biology & Philosophy*, 7. 1(1992), 69–76.

Sachs, Julius, 'Stoff und Form der Pflanzenorgane II', *Arbeiten des botanischen Instituts Würzburg*, 2(1882), 689–718.

Sauer, Tilman and Raphael Scholl, eds, *The Philosophy of Historical Case Studies* (Dordrecht: Springer 2016).

Schickore, Jutta, 'What Does History Matter to Philosophy of Science? The Concept of Replication and the Methodology of Experiments', *Journal of the Philosophy of History*, 5(2011), 513–532.

Schickore, Jutta, *About Method: Experimenters, Snake Venom, and the History of Writing Scientifically* (Chicago/London: University of Chicago Press, 2017).

Scholl, Raphael, 'Scenes from a Marriage: On the Confrontation Model of History and Philosophy of Science', *Journal of the Philosophy of History*, 12. 2(2018), 212–238.

Scholl, Raphael and Kärin Nickelsen, 'Discovery of Causal Mechanisms: Oxidative Phosphorylation and the Calvin-Benson Cycle', *History and Philosophy of the Life Sciences*, 37. 2(2015), 180–209.

Scholl, Raphael and Tim Räz, 'Towards a Methodology for Integrated History and Philosophy of Science', in *The Philosophy of Historical Case Studies*, ed. by Tilman Sauer and Raphael Scholl (Dordrecht: Springer, 2016), 69–91.

Schürch, Caterina, 'How Mechanisms Explain Interfield Cooperation: Biological-Chemical Study of Plant Growth Hormones in Utrecht and Pasadena, 1930–1938', *History and Philosophy of the Life Sciences* 39. 3(2017), article 16, 1–26.

Shapin, Steven and Simon Schaffer, *Leviathan and the Air-Pump: Hobbes, Boyle, and the Experimental Life*, paperback reissue, with a new introduction, originally published in 1985 (Princeton: Princeton University Press, 2011).

Stark, Peter, 'Das Reizleitungsproblem bei den Pflanzen im Lichte neuerer Erfahrungen', *Ergebnisse der Biologie*, 2(1927), 1–94.

Vicedo, Marga, 'Is the History of Science Relevant to the Philosophy of Science?' *PSA 1990: Proceedings of the Biennial Meeting of the Philosophy of Science Association*, 2(1992), 490–496.

Went, F.A.F.C., *Lehrbuch der Allgemeinen Botanik* (Jena: Gustav Fischer, 1933).

Wille, Robert-Jan, 'The Coproduction of Station Morphology and Agricultural Management in the Tropics: Transformations in Botany at the Botanical Garden at Buitenzorg, Java 1880–1904', in *New Perspectives on the History of Life Sciences and Agriculture*, Archimedes 40, ed. by Denise Phillips and Sharon Kingsland (Cham: Springer, 2015), 253–275.

Williams, L. Pearce, 'Should Philosophers Be Allowed to Write History?' *The British Journal for the Philosophy of Science*, 26. 3(1975), 241–253.

Wimsatt, William C., 'Reductive Explanation: A Functional Account', in *PSA 1974: Proceedings of the 1974 Biennial Meeting Philosophy of Science Association*, ed. by James van Evra (Dordrecht: Reidel, 1976), 671–710.

Wray, K. Brad, 'The Epistemic Significance of Collaborative Research', *Philosophy of Science*, 69. 1(2002), 150–168.

Zallen, Doris T., 'The Rockefeller Foundation and Spectroscopy Research: The Programs at Chicago and Utrecht', *Journal of the History of Biology*, 25. 1(1992), 67–89.

3   Narrative explanations in integrated History and Philosophy of Science

Claudia Cristalli

Introduction

'Narrative explanation' is a concept originally developed by analytic philosophers of history and only very recently applied to the Philosophy of Science.[1] In analytic Philosophy of History, narrative explanation has the role of bridging the gap between the analysis of individual 'historical' events and their attribution to a bigger process – namely, history itself.[2] The aim of this chapter is to expand on the recent work done on narrative in Philosophy of Science by considering how it may help provide a general framework for current models of scientific explanation. The originality of my approach lies in using concepts from the history of philosophy such as 'colligation' to expand the current place of 'narrative' in scientific explanation, thus providing a case for integrating History and Philosophy of Science. Narrative was introduced to account for explanation especially in the so-called historical sciences. 'Historical sciences' are those sciences, like palaeontology or geology, which study phenomena happening over a stretch of time too large to be studied experimentally. In this chapter, narrative is explored as a tool for understanding how explanation in science is historically laden and historically driven. Narrative structures provide the tools for the advancement of inquiry at least in the case of historical sciences. Since a satisfactory definition of narrative for the purposes of explanation has not yet been found, my analysis starts from some recognisable effects of narrative on inquiry. I see narrative as contributing to explanation in at least three ways: (1) narrative pulls together a great amount of facts; (2) narrative uses expectation and surprise and (3) narrative uses belief and doubt as steering wheels to navigate the amount of facts. I propose that 1–3 will help to build a general framework to understand different models of explanation. To make the discussion of models of explanation more concrete, I address two specific models: the experimental and the common cause explanation.

The first part of the chapter introduces the classic account of the ways in which experimental science provides explanations, as theorised by Carl G. Hempel and Paul Oppenheim (1965). The second part deals with Carol Cleland's criticism of Hempel and with her positive account of explanation. Cleland argues for the superiority of the historical 'common cause' principle over Hempel's law-based model for explanation. Metaphysical efforts to clarify what common cause is do not explain how common cause works, nor give it any superiority over the law-based model. The third part suggests that narrative explanation provides the context for both Cleland's historical explanations and Hempel's law-based explanations. The three features of narrative listed above are linked to the philosophers and scientists who first expounded them, namely William Whewell (1794–1866) and Charles S. Peirce (1839–1914). Via an analysis of these three features of narrative which are considered in the work of these figures, the History and Philosophy of scientific explanation come together and help inform our current ways of understanding of the role of explanation.

3.1 Experiment and generalisation

3.1.1 Experiment

In his 1948 'Studies in the logic of explanation', co-authored with Paul Oppenheim, Hempel defines scientific explanation as essentially the answer to the question 'Why?'. For Hempel, an answer to the question 'Why something is the case' involves the production of (1) antecedent conditions and (2) general laws. Under (1) we understand all description of circumstances and particular details, while (2) are the scientific laws that act as general premises of any scientific explanation (Hempel and Oppenheim 1965, p. 246). Both (1) and (2) have to be expressed propositionally, and the resulting explanation has the form of a logical deduction. This is why Hempel's model of explanation is also known as the 'deductive-nomological' account of explanation. Understanding explanation as a logical structure, Hempel can model explanation without the need to refer to physical causation, a notoriously problematic concept since the times of Hume. However, this move comes at a price. The price is the need for any scientific sentence to possess empirical content suitable for being tested experimentally. Only thus a conclusion reached via logical inference also possesses empirical value.

The experiment has thus the function of connecting a system of well-derived propositions with the reality they are talking about. A scientific proposition without experimental content would lack any possible connection with the world, and its scientific meaning could not be assessed. It would be impossible to distinguish between genuine scientific explanations and pseudo-explanations.[3]

3.1.2 Generalisation

The need for generality in a scientific explanation is intertwined with the need for a specific conception and use of the experiment. On the one hand, experiment gives empirical content to any general law. On the other hand, the role of the experiment is not to represent a particular phenomenon but rather any possible occurrence of that phenomenon. What is tested in any particular experiment is some general property of a natural phenomenon rather than its particular instance. As a consequence of this, laws never predict for the occurrence of particular event, but highlight those relations between events conforming to a general pattern. As Hempel puts it:

> ... all that a causal law asserts is that any event of a specific kind, i.e. any event having certain specific characteristics, is accompanied by another event which in turn has certain specified characteristics; for example, that in any event involving friction, heat is developed.
>
> (Hempel and Oppenheim 1965, p. 253)

Here, 'heat' and 'friction' are general phenomena which cannot be encountered as such. They are rather met as properties or effects of the interaction of individual bodies. The scientist is not interested in the individual bodies themselves but only in the possible, general effects of their interaction. Such effects can be measured and replicated, i.e. tested, in experiments.

The problem in having generality as an essential feature of explanation is that, while it makes sense for experimental science, it seems to do little justice to explanation in different contexts, as for instance history or the so-called 'historical sciences'. History and historical sciences share a commitment to investigate particular events, such as the Napoleonic wars or dinosaur extinction. In this context, the study of possible causes is not as interesting as trying to establish the actual (particular) causes.

Against the thesis of the radical difference in the objects of experimental and empirical sciences, Hempel maintains in his 1942 essay 'The Function of General Laws in History' that historical explanations too make use of general laws, or, better, 'general hypotheses'. Generality is indeed so essential to the historian's practice that it permeates all historical reconstruction:

> Even if a historian should propose to restrict his research to a 'pure description' of the past, without any attempt at offering explanations or statements about relevance and determination, he would continually have to make use of general laws. For the object of his studies would be the past – forever inaccessible to his direct examination. He would have to establish his knowledge by indirect methods: by the use of universal hypotheses which connect his present data with those past events.
>
> (Hempel 1965, p. 243)

In the end, Hempel presents his notion of general hypothesis in history as a contribution against the artificial division between descriptive and theory-laden constructions. Despite his efforts, however, it remains apparent that the generalisations present in history and in historical sciences – such as the 'universal hypothesis' that we have access to past events – are not the aim of historical explanation. In history, explanation is no longer deductive, because it does not follow logically from a set of premises, nor is it nomological, because laws have, in this framework, only an ancillary role. Eventually, the effort to understand historical explanation with the same epistemic tools developed for experimental science may have a rather counter-productive effect, namely that historical explanation falls short of the standards of experimental science. Instead of being examined in its own right, historical explanation is understood as a defective approximation of the standards of experimental science.

3.2 Explanation in historical sciences: the common cause account

Carl Hempel's account of historical explanation influenced the twentieth century debate on scientific explanation so deeply that almost every discussion of explanation grapples with it.[4] Among this vast literature, Carol Cleland stands out not only as a defender of historical sciences' methodology, but also as a proponent of their epistemic superiority over empirical experimental science.[5] I firstly define explanation in historical sciences and then move on to a closer examination of the common cause account.

3.2.1 Explanation in historical sciences

The first step of Cleland's argument is that historical sciences' models of explanation shall not be judged against the standards of experimental science; instead, they have to be assessed in their own right.

According to Cleland, historical sciences – such as palaeontology, geology, and evolutionary biology – are not interested in describing the general properties of phenomena, whose individual occurrence is irrelevant. Historical sciences are interested in particular events of natural history. Historical sciences thus share with history the interest in what *actually* happened. When talking about the future, historical sciences predict *possible* outcomes: they do not tell what will happen in any case but what might happen in specific cases. Ultimately, the output of historical sciences is not another law, but rather a more or less plausible reconstruction of a particular event.

If, however, the kind of explanation developed by experimental sciences is considered paradigmatic of science as such, generalisation and experiment become the universal requirements for any practice that wants to be scientific. Historical sciences however cannot but fall short of such requirements. The identification of scientific explanation with experimental explanation has a negative impact on the prestige of historical sciences, and this contributes to the marginalisation of historical sciences within the scientific community (Cleland 2009, p. 45).

Cleland argues that explanation in historical sciences is supported by a metaphysically robust account of causation, which would do justice to the research practice of historical scientists. While an experimental scientist may want to test different possible implications of a unique hypothesis, historical scientists are often dealing with a variety of plausible hypotheses for a particular, puzzling event (Cleland 2001, p. 989). Scientists have to pick one hypothesis on the basis of scarce evidence and the plausibility of the story it tells. Intuitively, one way of doing so is to look for the hypothesis which explains the greatest part of the available evidence in a smooth and coherent way. This amounts to postulating a *common cause* for the available evidence.

3.2.2 The common cause account of explanation

The second step of Cleland's argument is to argue that the common cause account of explanation is not just an equally viable explanatory strategy, but in fact it is a more rigorous and a safer ground for scientific explanation.

Within historical sciences, to postulate a common cause does not mean *deducing* it from some general characters of what 'causes' are; rather, a common cause is inferred from the available evidence because of its power to weave such evidence into a coherent story. As Cleland (2011, p. 569) states:'The principle of the common cause represents an epistemological conjecture about the conditions under which a certain pattern of causation may be non-deductively inferred.'

Accordingly, the common cause principle provides precisely the epistemic structure that can make sense of hypothesis formation and the fitting together of the available evidence in a non-deductive fashion. A common cause does for the historical sciences what Hempel's model of explanation does for experimental sciences. Hempel could give a logical justification of his model (Hempel 1943). What, then, would the justification of the common cause principle be? Cleland's move is to ground the common cause principle in one of the fundamental physical dimensions of our world: time.

Time, as a physical dimension, has an oriented structure, which can be visualised by an arrow (Reichenbach 1956). The transition from possibility to actuality – the becoming part of the future – involves a radical pruning of alternatives, and no *a priori*, purely logical method can enable us to see which ones will be cut off. This means that, when looking for causation in nature, the past is observed to be *overdetermined* over the future: while we can usually individuate the causes of some fact F, we cannot foresee all the possible effects of the same fact F. This is why, Cleland (2011, pp. 470–1) explains, it is relatively simple to determine from the analysis of scant traces that an eruption occurred millions of years ago, but it is almost impossible to predict from the multitude of available data when Vesuvius will erupt again.

The fact that we find ourselves much better at inferring past causes from present effects than at inferring future effects from present causes is called *asymmetry of overdetermination* (AO). AO states 'that events in our universe are causally connected in time in an asymmetrical manner' (Cleland 2011, p. 570). Cleland borrows AO from David Lewis (1979), and she applies it to physical processes at a rather general level. More precisely, Cleland (2011, p. 571) maintains that 'all physical phenomena above the quantum level are subject to the asymmetry of overdetermination.' Accordingly, it would seem that for Cleland this principle has the same strength and scope of any law of classical physics. For Lewis however, who is in this respect openly under Frank P. Ramsey's wing, natural laws are named such only by approximation: 'proper' laws are essentially deductive systems, i.e., formal constructs.[6] According to Lewis, AO is not some feature of the world, but just a general proposition that is true in our world but may well not be true in other worlds. It has no necessity in it, as all necessity is purely logical. From this premise, Lewis (1973, p. 73) defines scientific practice as the best possible approximation to a deductive system which is true about an actual world *i*: 'Our scientific theorizing is an attempt to approximate, as best as we can, the true deductive systems with the best combination of simplicity and strength.'

The fact that the laws of nature are contingent does not make them random. In fact, in a later paper, Lewis (1979, pp. 474–5) uses the asymmetry of overdetermination to illustrate the implausibility of miracles, which would be events entirely independent from the past. However, what matters for us is that natural laws are for Lewis imperfect approximations to 'the true deductive systems' of logic. Despite using Lewis in support of the common cause principle, Cleland would not buy into the idea that this same principle depends on some (imperfect) deductive system. In fact, her whole philosophical project, which aims at defending historical explanation from claims that it does not fit the deductive model of scientific explanation, demonstrates the relevance of contingent and empirical findings to the scientists' work. If the common cause principle means to be an alternative to deductive systems of justification, Cleland may want to support it with a different argument.

Granted that the common cause account of explanation works in historical sciences, its explanation in term of the Lewisian AO principle is not satisfactory. Moreover, AO would not undermine the traditional account of how experimental science works, nor put historical science in a position of epistemic privilege. As I show in the next section, it is more productive to embed the common cause account of explanation in a different framework than the *a priori* and deductive structure of Lewis' metaphysics.

A productive framework for the common cause account of explanation is one which shows us how common cause explanations work. What stimulates us to the search of a common cause, and how do we proceed in this search? What logic do we employ in looking for a common cause? Could it eventually be used to promote a less mythical understanding of experimental science? These questions have been the bread and butter of philosophers of history. Today, they begin to be unpacked by philosophers of science.[7] The idea currently on the table is to look at narrative and at its structure to understand scientific explanation. Incidentally, Cleland too touches upon the question of what drives the construction of an explanation when discussing how historical scientists work:

> the more improbable an association among a collection of traces seems the more psychologically appealing the claim that it is the product of a common cause. This helps to explain why historical natural scientists have a tendency to focus their investigations on what seems in light of

their background beliefs to be *the most unlikely* (and hence *puzzling*) correlations or similarities among contemporary phenomena.

(Cleland 2011, p. 569, emphasis added)

In looking for explanation, puzzlement and plausibility go hand in hand. In any kind of research, it is crucial *which* puzzle (over many possible ones) will be chosen as worthy of our attention. The more an event clashes with the background of our currently held beliefs, the more it compels us to look into it, and the more such research may be rewarding. While deductive systems do not leave much space for puzzlement, it is often the aim of narratives to arouse, cultivate and direct puzzlement and surprise. The very structure of narrative seems to be built on developing an awareness of how different pieces may – or may not – fit together.

3.3 Narrative explanation

In this section, I focus on the explanatory power of narrative. Narrative explanations have long been the object of study of analytical philosophers of history. Because of this tradition, it is important to be clear at the outset on what, in the present context, narrative explanation *is not*:

a   It is not an 'Ideal Chronicle', that is, 'a narrative consisting of all the true statements that could be made about the world right from Adam and Eve down to the present' (Danto 1985, p. 152, quoted in Ankersmit 2009, p. 202). Such an account implies a purely linguistic idea of narrative and an idea of explanation as 'rational reconstruction'. Both conditions are highly undesirable because they add the problem of connecting the narrative back to what it should explain – a problem known to philosophers as 'the problem of reference'.

b   It is not a 'verbal fiction, the contents of which are as much *invented* as *found* and the forms of which have more in common with their counterparts in literature than with those in science' (White 1978, p. 82, original emphasis). It is good to acknowledge that narrative explanation is not a pile of matters of fact; still, the blind sorting of its content into 'invented' and 'found' perpetrates the positivist division between 'interpretation' and 'fact'.[8] As for the forms of narrative, we want something simpler than what is in use in literature and closer to the scientists' practice.

c   It is not a counterfactual history of what could have happened and did not. A counterfactual history is a 'what if' history, whose purpose however seems to be to enlarge – rather than help direct – the pool of available hypotheses.[9]

After all these caveats, it is legitimate to ask again: What is narrative explanation? Geoffrey Hawthorn maintains that narrative is a kind of novel in that it explains in virtue of the setting it creates, and through it, the reader's faculty of judgement is trained and nurtured.[10] Despite agreeing on the importance of judgement, Hawthorn's exclusive attention to details risks transforming narrative into a series of exemplary cases where no generalisation is possible. Instead, narrative explanations work also in virtue of certain explanatory structures. The following section looks at those crucial structures of narrative that – I claim – can be found in experimental and historical explanations alike.

3.3.1 The explanatory structure of narrative

Usually, narrative theorists distinguish between *narrative* and *chronicle*. The difference between a chronicle and a narrative is that a narrative offers relational elements that join together a sequence of events, whereas a chronicle just lists them (Morgan 2017b). Both narratives and chronicles are constituted by a series of events happening in time, however we understand narratives as conveying more than chronological order. But what is this 'more'? We are provided here with a puzzling fact – narratives work in conveying explanations – but we do not know *how* they work. Paul Roth's suggestion is to look at narrative in terms of its conventions and at its structures:

> Rather than attempting to look beyond conventions of narrative, perhaps the proper strategy is to insist that *it is these very conventions which do the work of explanation*. In this respect, narrative conventions are constitutive of historical practice and determinative of historical explanation. The suggestion I am entertaining is that the *conventions define the possibilities of explanation as well*.
>
> (Roth 1989, p. 460, emphasis added)

Roth is suggesting here that narrative structures are not only doing the epistemic work of fitting the collected evidence into an explanation, but that they are the condition of possibility of narrative explanation as such. If this is true, it becomes crucial to determine the structures of narrative and which structures are most promising for scientific narrative.

Mary Morgan (2017a, p. 2) elaborates on the idea that narrative structures *do the work* of explanation even beyond theories: while theories appear too detached from particular facts to really account for them, narratives may provide 'a natural form for bringing related elements into order or creating order out of dis-ordered materials.' My proposal of narrative structures of explanation goes directly at unpacking what Morgan describes as a 'natural form'. John Beatty points in this direction when he suggests a shift in perspective between narratives and more traditional forms of explanation:

> Instead of asking whether narratives measure up to other forms of representation and explanation, we should consider what they do especially well, what they may do better than other forms of representation and explanation, and why science would be impoverished without them.
>
> (Beatty 2017, p. 41)

So, what do narratives do especially well? I claim that narratives are particularly good at: (1) knitting particular facts together in a way that theories cannot; (2) exploring contingent possibilities and giving contingent explanations for contingent events; and (3) at the same time attaining a kind of generality via the use of general structures. We now turn towards an analysis of some of those general structures. This analysis relies on the introduction of two important yet overlooked figures in the History of Science and of the Philosophy of Science, namely the Cambridge polymath William Whewell (1794–1866) and the American logician, scientist and philosopher Charles S. Peirce (1839–1914). Whewell's expertise ran from mathematics and meteorology to the Philosophy of Science. The influence of Kant on his ideas prompted him to critically reformulate Bacon's method for the advancement of science. Peirce too was deeply interested in the advancement of science and in the development of methods for achieving this aim. Both were writing before the (now classic) distinction between context of discovery and context of explanation[11] and the subsequent confinement of Philosophy of Science to the context of explanation only. When, in 1958, Norwood Russell Hanson (1958) argued for the possibility of a logic of discovery, some philosophers of science started looking at the History of Science as a great source for the study of discovery. During the nineteenth century however, philosophers talking about science would look at its historical development as well as at its more recent outputs, with the conviction that history had many lessons to teach the philosopher and the practitioner of science alike. Sharing this conviction, I propose that we look forward at narrative explanation by looking back at Whewell and Peirce.

3.3.2 Colligation

'Colligation' is a term famous to philosophers of science thanks to William Whewell (1794–1866), who also coined the more popular term 'scientist'.[12] Whewell firmly believed in the integration of History and Philosophy of Science and in the role of history for the advancement of science. In his 1858 *Novum Organon Renovatum*, Whewell (1858a, p. iii) proposes to renew Francis Bacon's project from the vantage point of history: 'Bacon could only divine how sciences might be constructed; we can trace, in their history, how their construction has taken place.'

Whewell's historical understanding of the development of science makes him aware of the social and institutional elements involved in the production of science and of the pluralism of methods across the sciences.[13] From an epistemological perspective, Whewell integrates induction with general ideas. While the method of research is inductive, its guidance and support come from general ideas. Colligation, being the central step of Whewell's model for induction, can be seen as his explanation of how we get from particular facts to general laws. The classical example of colligation is Kepler's induction from Tycho Brahe's observational data on the positions of planets to the idea that planetary orbits are elliptical, since it involves finding a general conception that fits the existing data. Once colligated, the particular instances exhibit something which was not evident before such an action was performed. This *something*, according to Morgan (2017b, p. 89), is the relation they have to each other, which may be embodied by a 'guiding conception' or a 'categorization schema': 'I use colligation to capture the way a scientist both brings together, and assembles, a set of similar elements framed under some overall guiding conception, or categorization schema.'

Relations are extremely rich epistemically in that they enable our understanding of particular objects at the more general level of structures. The generality they provide is not alien to the particular data or events, but rather arises from their own combination. The language used here by Morgan is reminiscent of Kant, however she does not discuss Whewell's Kantianism, nor whether it could influence the role of colligation in her account of narrative (Ducheyne 2011). Where Whewell makes colligation a 'faculty of thought', Morgan does not establish a hierarchy between bottom-up relations and top-down categorising schemata. The two chief merits of her contribution consist (1) in regarding colligation as a narrative structure and (2) in opening up the analysis of colligation beyond cases from experimental science, going into sociology, history, and art. As an example of colligation in art Morgan brings Bruegel the Elder's painting 'Children's Games', dated 1560.[14] For Morgan (2017b, p. 88), Bruegel's painting exemplifies colligation because it offers 'closely observed bits of life fitted under a conceptual title.' My interest is here in the process of observation itself and on the mutual interaction between the title (which stands for Whewell's 'ideas') and the 'facts' represented. In pursuing this, I will push Morgan's analysis a bit further.

The painting, in the typical Nordic style of the sixteenth century, uses perspective not to focus the viewer's gaze but to distribute it around the space, putting its subject on display as if in a cabinet of curiosities.[15] To this effect, only geometric perspective is used, which keeps the figures in the background as focused as the ones in the foreground. The title 'Children's Games' comprehends the many subjects of the painting as the simplest interpretative hypothesis. The title functions as a springboard into the painting itself and into the observation of the multitude of the more than two hundred figures it depicts. Amy Orrock (2012, p. 2), who looked at the practices of the painting's fruition in its own times, reports that 'Bruegel's paintings were most often displayed in private social spaces, such as dining rooms.' They would then be exposed to a public selected by invitation, who could have accessed them in the social context of a dinner. As a result, 'Bruegel's complex panels would have functioned as conversation pieces, providing focus for debate during gatherings of like-minded individuals. Van Mander's anecdotes … are evocative of a culture in which looking was an active, and, for some, competitive sport' (Orrock 2012, p. 3). Reading Bruegel's paintings involves therefore running one's eye back and forth on the many little figures populating it in order to reason by analogy and association – a task where creativity is enhanced by experience. Because of the continuous interplay between observation and interpretation, any hypothesis that arises from looking at Bruegel's painting will influence the way the painting is seen. Such a hypothesis will be perceived, at least as long as the interpretation holds, as a 'fact' of the painting itself. Any 'conceptual title' will, as Whewell (1858b, p. 49) suggests, become a 'fact' from which future observation will start off: 'Theories become Facts, by becoming certain and familiar: and thus, as our knowledge becomes more sure and more extensive, we are constantly transferring to the class of facts, opinions which were at first regarded as theories.'

By attributing a fluid nature to the labels 'fact' and 'theory', Whewell can be (anachronistically) described as being 'perspectivist' about science. What at a given time is a rigid division between fact and theory does not hold at a later stage: the theory may be discarded or it might be incorporated into a much broader theoretical framework, where it is actually treated as a fact; the fact, pressed by further analysis, may show itself as the product of false theoretical assumptions (Whewell 1858b, p. 50).

Eventually, colligation is a fundamental narrative structure of explanation because it accounts for the process of induction in two directions: from the data to the theory, and from the theory to the data. Colligation makes the role of analogy and association explicit, as well as the importance of perspective in any explanation. Colligation is about putting together all the relevant evidence into a meaningful shape, but it is not an automatic process. In opposition to Francis Bacon, Whewell (1858a, p. v) did not claim to have the recipe for an 'Art of Discovery', since no art can 'enable all men to construct Scientific Truths as a pair of compasses enables all men to construct exact circles.' More modestly, colligation is one among the methods which bring discoveries about.

3.3.3 Expectation and surprise

If correlations and similarities among phenomena emerge from the narrative structure of colligation, it is the *puzzlement* arising from an event which does not fit in the scientists' *background belief* that prompts the colligatory process to take place.[16] Puzzlement and background belief are captured by the concepts of 'surprise' and 'expectations'; their role in scientific inquiry was first theorised by a very careful reader of Whewell – Charles Sanders Peirce (1839–1914). Peirce was also a man of sciences with vast interests, but, contrary to Whewell, he never held a permanent academic position. Peirce is best known today as the first American philosopher and the founder of American pragmatism. Together with 'expectation' and 'surprise', Peirce (1992a) used the couple of concepts, 'belief' and 'doubt', to describe the process of scientific inquiry. They are not entirely technical terms. In fact, part of their strength derives from their appeal to common sense. In the following, the two couples expectation/surprise and belief/doubt will be used almost as key structures of narrative explanations.

For Peirce, expectation and surprise are not just psychological statuses. They are assumed to be logical and epistemic structures of understanding. Expected events are habitual events: surprising events may trigger doubt. Doubt is the spur of inquiry. To get rid of doubt, we look for knowledge that could confirm our expectations in the long run. Doubt is indeed so crucial for Peirce's epistemic theory that his first philosophical publications – from 1868 to 1878 – are also known as 'anti-Cartesian essays'. This is due to Peirce's insistent criticism of the idea that inquiry may start with an absolute doubt. Such a doubt, Peirce claims, is nothing but a 'paper doubt', it has no *real*, prodding effect over inquiry: 'the mere putting of a proposition in the interrogative form does not stimulate the mind to any struggle after belief. There must be a real and living doubt, and without this all discussion is idle' (Peirce 1992a, p. 115).

In contrast, a 'living doubt' arises when something disappoints our belief, i.e. our expectation. A belief opportunely generalised is a 'habit of mind,' also called '*guiding principle* of inference'. The habit of mind 'determines us, from given premises, to draw one inference rather than another' (Peirce 1992a, p. 112). The inference Peirce has in mind can be drawn from one experimental instance to the whole class of objects possessing a certain property, as in Hempel's account of experimental explanation, or can be drawn from the colligation of many particular facts into an overarching hypothesis.

Peirce had started distinguishing between deductive and inductive or hypothetical inferences already in his Harvard Lecture of 1865 and in his Lowell Lectures of 1866, both on the logic of science (Peirce 1982a, 1982b). In those lectures, Peirce shows the greatest interest in induction. He deeply engages with syllogism, proving that the two 'oblique' models of inference are truly non-deductive because they cannot be reduced to a deductive syllogism without the addition of a fourth proposition. He therefore establishes *formally*, from the point of view of the relation of arguments and without reference to their content, that induction has an autonomous logical status, distinct from that of deduction. As Peirce (1992a, p. 112) states:

> The object of reasoning is to find out, from the consideration of what we already know, something else which we do not know. Consequently, reasoning is good if it be such as to give a true conclusion from true premises, and not otherwise. Thus, the question of its validity is purely one of fact and not of thinking.

The factual part of explanation, that in Hempel's account was strictly connected with experiment, is for Peirce already embedded in the proper kind of inference used in science. According to Peirce, explanation in science cannot be attained by deduction only (although deduction still maintains an important place in reasoning). More importantly, for our concerns in the integration of History and Philosophy of Science, logic is, in Peirce's perspective, an epistemic tool which is itself contingent on the current state of knowledge. It can and must be used in scientific reasoning, but only with the insight that it can and must be reformed as our science evolves. Indeed, any major progress in science *is* a progress in its method, and therefore a reform in its logic. Logic must be rooted in the History of Science:

> Every work of science great enough to be remembered for a few genera-
> tions affords some exemplifications of the defective state of the art of
> reasoning of the time when it was written; and each chief step in science
> has been a lesson in logic.
>
> (Peirce 1992a, p. 111)

This position calls to mind Whewell's perspectival understanding of the relation between facts and theory. Both Peirce and Whewell reach this epistemic stance because of their deep interest in the History of Science joined with a lively practice of some of its branches. Both see the logic of science as a contingent affair, which derives its meaning and the stimuli for its growth from the 'facts' and problems of the sciences of the time. Expectation and surprise, as 'living principles of inference', appear as an invaluable resource for the identification of problems in the first place: they are the very starting point of inquiry. Colligation then is the structure that helps to build hypotheses out of a multitude of data. Together, expectation, surprise and colligation constitute a – partial and perfectible – account of how narrative explanations operate in science.

Conclusion

This chapter proposes narrative as a general framework for scientific expla-nation. Such a framework does not deny the important differences of existing models of explanation: rather it aims to find tools to orientate the use of those existing models of explanations. Questions more related to the process of inquiry, such as how do we pick our hypotheses, or how do we decide which problems catch the scientist's attention, may be also considered within a narrative framework. Fundamentally, the narrative framework of explana-tion undermines the idea of a privileged model of scientific explanation. In this case, experimental sciences need not be given higher explanatory value than historical sciences.

Using a narrative model of explanation does not mean to adopt the idea that narrative structures should be reduced to linguistic structures. The nar-rative model is compatible with certain kinds of counterfactual models of explanation; however, the narrative model does not imply that narrative explanation works in virtue of it providing a counterfactual history. Instead, the contribution of narrative is seen in (1) understanding the dynamic between facts and theories or particular instances and general laws, (2) enabling a perspective understanding of the present in virtue of the past experiences and the future expectations, and (3) giving a general account for how certain events attract the attention of the scientific community with the notion of surprise.

The narrative structures 1–3 proposed here are just a first step towards a possible narrative interpretation of explanation in integrated History and Philosophy of Science. More needs to be done to make narrative a truly comprehensive framework to understand explanation. We need to ascertain how colligation on the one side and expectation and surprise on the other side are actually cultivated in the practice of science. Moreover, theoretical analysis is likely to reveal more structures playing a role in narrative explanation. As presented here, narrative explanation provides a starting point for dealing with conflicting methodologies in History and Philosophy of Science, as well as contributing more specifically to the debate on explanation in the sciences.

Notes

1 For a classical analytical account of narratives in philosophy of history see Danto (1965). For recent application of narrative to philosophy of science see Mary Morgan (2017a).
2 For a comprehensive introduction on philosophy of history see Lemon (2003).
3 Indeed, there has been a lot of work on logic in this vein since the 1960s. See for instance Norton (2015); and Norton (2010). Thanks to Jono Spring for pointing this out to me.
4 For an introductory yet comprehensive account of the origins of the deductive-nomological account of explanation and of its problems, see Woodward (2017)
5 See Carol Cleland (2001, 2009, 2011).
6 David Lewis (1973), p. 73: 'We can restate Ramsey's 1928 theory of lawhood as follows: a contingent generalization is a law of nature if and only if it appears as a theorem (or axiom) in each of the true deductive systems that achieves a best combination of simplicity and strength. ... lawhood is a contingent property. A generalization may be true as a law in one world, and true but not as a law in another [world].'
7 Ankersmit (2009), p. 200. Ankersmit attributes this question to Haskell Fain (1970), but without mentioning his answer: 'Yet, it is the story-line that makes the relationship between historical facts intelligible, that dictates how facts should be described, which facts selected as relevant and which omitted [...]' (p. 255).
8 Against this view, see Edward H. Carr (1964), pp. 29–30.
9 Steve Fuller (2008) imagines counterfactuals as an 'exercise' that enables a dialogue with the past, but it is a dialogue in which we have the first and the last word. Gregory Radick (2008) instead uses counterfactual narration to give voice to unrealised possibilities of the past. Finally, Steven French (2008) doubts the use of counterfactual histories, in that it is hard that they represent 'genuine possibilities' – namely, a possibility that would stand some serious chances of having been actual.
10 Geoffrey Hawthorn (1991), p. xi: '...the resources that we need to make such judgements, I argue, are given in the details of particular cases.' The hope that detailed narratives will turn the reader of science into a witness dates back to Robert Boyle. See Steven Shapin (1984, pp. 490–491).
11 This distinction was originally introduced by Hans Reichenbach (1938).
12 Whewell was Master at Trinity College, Cambridge, where he had been studying and living since 1812. Whewell's interests were broad: professor of mineralogy and philosophy, Whewell also did important work for the prevention of the Thames' tides and was conversant with the basic ideas of most sciences of his time. In philosophy, Whewell is an original and insightful follower of Kant.

13 Whewell (1858a), p. iii: 'it would be worthy undertaking to determine the machinery, intellectual, social, and material, by which human knowledge can best be augmented.' See also Whewell (1858b), p. 48: 'Thus we are often told that such a thing is *a Fact*; A FACT and not a Theory [...]. [...] we must ask to whom is it a Fact? What habits of thought, what previous information, what Ideas does it imply, to conceive the Fact as a Fact?'. Books 2–5 of Whewell (1858b) v.1, analyse in detail the different methodological standpoints of the different sciences.
14 Pieter Bruegel, 'Children's games' (1560, Kunsthistorisches Museum in Vienna).
15 Cabinets of curiosities or *Wunderkammern* (literally, 'chambers of wonders') were rooms where sixteenth century notables would display 'weird' or 'suggestive' objects they had collected. These items (including natural objects and artifacts) were juxtaposed in creative ways to stimulate the visitor's own analogies and associations.
16 Cleland (2011), p. 569: 'historical natural scientists have a tendency to focus their investigations on what seems in light of their background beliefs to be the most unlikely (and hence puzzling) correlations or similarities among contemporary phenomena.' On this topic see Caterina Schürch, 'What to Ask and How to Answer: Investigating Research Planning in General Physiology', in this volume.

Bibliography

Ankersmit, F., 'Narrative and Interpretation', in *A Companion to the Philosophy of History and Historiography*, ed. by Aviezer Tucke (Chichester: John Wiley and Sons, 2009), 199–208.

Beatty, J., 'Narrative Possibility and Narrative Explanation', *Studies in History and Philosophy of Science*, 62(2017), 31–41.

Carr, E.H., *What is History?* (London: Penguin 1964, 2nd ed.).

Cleland, C., 'Historical Science, Experimental Science, and the Scientific Method', *Geology*, 29. 11(2001), 987–990.

Cleland, C., 'Philosophical Issues in Nat Hist and Its Historiography', in *A Companion to the Philosophy of History and Historiography*, ed. by Aviezer Tucker (Chichester: John Wiley and Sons, 2009), 44–62.

Cleland, C., 'Prediction and Explanation in Historical Natural Sciences', *British Journal of Philosophy of Science*, 62(2011), 551–582.

Danto, A., *Narration and Knowledge* (New York: Columbia University Press, 1985).

Ducheyne, S., 'Kant and Whewell on Bridging Principles between Metaphysics and Science', *Kant-Studien*, 102. 1(2011), 22–45.

Fain, H., *Between Philosophy and History: the resurrection of speculative philosophy of history within the analytic tradition* (Princeton: Princeton University Press, 1970).

French, S., 'Genuine Possibilities in the Scientific Past and How to Spot Them', *Isis*, 99(2008), 568–576.

Fuller, S. 'The Normative Turn: Counterfactuals and a Philosophical Historiography of Science', *Isis*, 99(2008), 576–584.

Hanson, N.R., *Patterns of Discovery* (Cambridge: Cambridge University Press, 1958).

Hawthorn, G., *Plausible Worlds: Possibility and Understanding in History and the Social Sciences* (Cambridge: Cambridge University Press, 1991).

Hempel, C.G. and P. Oppenheim, 'Studies in the Logic of Explanation', in *Aspects of Scientific Explanation and other Essays in the Philosophy of Science*, ed. by Carl G. Hempel (New York: The Free Press, 1965), 245–294.

Hempel, C.G., 'The Function of General Laws in History', in *Aspects of Scientific Explanation and other Essays in the Philosophy of Science*, ed. by Carl G. Hempel (New York: The Free Press, 1965), 231–243.

Hempel, C.G., 'A Purely Syntactical Definition of Confirmation', *The Journal of Symbolic Logic*, 8(1943), 122–143.

Lemon, M.C., *Philosophy of History* (London: Routledge, 2003).

Lewis, D., *Counterfactuals* (Cambridge: Harvard University Press, 1973).

Lewis, D., 'Counterfactual Dependence and Time's Arrow', *Nous*, 13. 4(1979), 455–476.

Morgan, M.S., 'Narrative Science and Narrative Knowing. Introduction to Special Issue on Narrative Science', *Studies in History and Philosophy of Science*, 62 (2017a), 1–5.

Morgan, M., 'Narrative Ordering and Explanation', *Studies in History and Philosophy of Science*, 62(2017b), 86–97.

Norton, J.D., 'Replicability of Experiment', *Theoria*, 30. 2(2015), 229–248.

Norton, J.D., 'There Are No Universal Rules for Induction', *Philosophy of Science*, 77. 5(2010), 765–777.

Orrock, A., 'Homo ludens: Pieter Bruegel's Children's Games and the Humanist Educators', *Journal of Historians of Netherlandish Art*, 4. 2(2012), 1–21.

Peirce, C.S., 'The Fixation of Belief', in *The Essential Peirce: Selected Philosophical Writings*, ed. by Nathan Houser and Christian Kloesel (Bloomington: Indiana University Press, 1992a), I, 109–123.

Peirce, C., 'On The Logic of Science [Harvard Lectures of 1865]' in *Writings of Charles S. Peirce. A Chronological Edition*, v. 1, ed. by Max H. Fisch et al., (Bloomington: Indiana University Press, 1982a), 161–302.

Peirce, C., 'The Logic of Science; or, Induction and Hypothesis [Lowell Lectures of 1866]', in *Writings of Charles S. Peirce. A Chronological Edition*, v. 1, ed. by Max H. Fisch et al., (Bloomington: Indiana University Press, 1982b), 357–504.

Radick, G., 'Introduction: Why What If?', *Isis*, 99(2008), 547–551.

Reichenbach, H., *Experience and Prediction* (Chicago: University of Chicago Press, 1938).

Reichenbach, H., *The Direction of Time* (Berkeley: University of California, 1956).

Roth, P., 'How Narratives Explain', *Social Research*, 56. 2(1989), 449–478.

Shapin, S., 'Pump and Circumstance: Robert Boyle's Literary Technology', *Social Studies of Science*, 14. 4(1984), 481–520.

Whewell, W., *Novum Organon Renovatum* (London: John W. Parker & Son, 1858, 3rd ed.). First published as *The Philosophy of the Inductive Sciences, Founded Upon Their History* (London: John W. Parker, 1840).

Whewell, W., *The History of Scientific Ideas* (Cambridge: Cambridge University Press, 1858).

White, H., *Tropics of Discourse* (Baltimore MD: Johns Hopkins University Press, 1978).

Woodward, J., 'Scientific Explanation', *The Stanford Encyclopedia of Philosophy* (Fall 2017 Edition), Edward N. Zalta (ed.), https://plato.stanford.edu/archives/fall2017/entries/scientific-explanation/.

4   Is a normative historically oriented Philosophy of Science possible?

A new horizon for integrated History and Philosophy of Science (iHPS)

Eugenio Petrovich

Introduction

In 1973, Ronald Giere famously framed the question of the relation between History and Philosophy of Science under the metaphor of *marriage*. 'History and Philosophy of Science: Intimate Relationship or Marriage of Convenience?' (Giere 1973), he asked in an important review of the debates taking place in the 1960s around the so-called historical turn in Philosophy of Science. As it is well known, the publication of *The Structure of Scientific Revolutions* by Thomas Kuhn, as well as other influential works by Norwood Hanson (1965), Mary Hesse (2005), and Paul Feyerabend (1962) fuelled the discussion between historians and philosophers of science. Was it possible to integrate History and Philosophy of Science? And if so, how?

Perhaps these questions are even more crucial today than in the 1960s, since the intellectual landscape around the integrated History and Philosophy of Science (iHPS) is nowadays filled with strong competitors. From esoteric formal epistemology to post-modernist Science and Technology Studies (STS), the disciplines studying science have considerably thrived in the last decades. In this context, iHPS cannot postpone the crucial question about its identity, mission and theoretical foundations. The aim of this chapter is to deal with one of the core theoretical problems within iHPS, namely how History and Philosophy of science should be integrated into iHPS. After introducing three key and distinct positions in the debates surrounding the *Structure*, I will propose a new horizon where the integration between the two could take place today: the arena of Science Policy. With this chapter, I claim that the Philosophy of Science Policy would allow a new iHPS to stand out today, both as a research programme, and as a call for action.

The first section of the chapter sums up three different views on the relationship between History and Philosophy of Science: the Kuhnian, the neo-positivist and the Popperian. I argue that these approaches form a Hegelian triad where Kuhn represents the thesis, neo-positivism the anti-thesis, and Popper the synthesis. I then focus on Popper's position, which I call a 'normative historically oriented iHPS', and I explain its distinctive logic, which I dub 'exemplary logic'. In the second section, I show how the same logic shapes the *Science Policy* discourse by analysing three different examples of Science Policy production: Vannevar Bush's report *Science, the Endless Frontier*, the so-called Triple Helix model proposed by Henry Etzkowitz and, lastly, a Science Policy document by the European Research Council. In the third and final section, I outline the research agenda of the Philosophy of Science Policy (PSP), a research programme that I propose as a new synthesis between History and Philosophy of Science, where the Popperian approach to iHPS is combined with a focus on Science Policy issues.

4.1 Three ways of conceiving the relation between History and Philosophy of Science

In this section I introduce three different views concerning the relationship between History of Science and Philosophy of Science: the Kuhnian, the neo-positivist and the Popperian. These positions were proposed during the discussion about the historical turn in Philosophy of Science in the 1960s and 1970s. I focus on the theoretical tenets of each of the positions, and the logical relations between them. I claim that all the arguments can be aligned with the classic Hegelian logic of thesis-antithesis-synthesis.[1]

4.1.1 Kuhn's empiricist model

The first position I introduce is the one developed by Kuhn in the first chapter of *The Structure*, which was tellingly entitled 'A Role for History'. The chapter ends with a statement where Kuhn (1970b, p. 9) clearly describes the optimal relation between historical reconstruction and philosophical theory: 'How could history of science fail to be a source of phenomena to which theories about knowledge may legitimately be asked to apply?'

The key words in the quote above are 'phenomena' and 'theories'. Kuhn's core idea is that History of Science provides the phenomena that theories of science (read Philosophy of Science) should account for. Hence, History of Science and Philosophy of Science stand in the same relationship as evidence and theory do. There is a clear division of duties between the historian and the philosopher of science, and their burden of proof is clearly separated. The historian of science collects and recounts a specific class of phenomena (phenomena that specifically regard the development of science), whereas the philosopher of science should explain them via a theory of scientific change. Historical facts represent the test of philosophical theories, which are therefore meant to be *empirical* theories about a specific class of phenomena. Within the Kuhnian model of the relationship between History and Philosophy of Science, the testing function of historical facts is guaranteed by their *independence* from philosophical theory, so that the direction of testing goes from Philosophy of Science to History of Science and not the other way around.[2] Here we can define a Hegelian triad where the Kuhnian empiricist model is the thesis. The thesis can be stated as follows: the relation between History and Philosophy of Science on the one hand, and evidence and theory, on the other, is the same. Specifically, the function of History of Science is to provide the evidence that Philosophy of Science aims at explaining by a theory.

4.1.2 The neo-positivist dualistic model and the normativity issue

In the context of neo-positivist philosophy of the 1960s, Kuhn's Philosophy of Science was highly contested. A detailed reconstruction of all the objections raised against *The Structure* lies outside the scope of the present chapter.[3] What is of interest here is how neo-positivists criticised the Kuhnian model of the relationship between History and Philosophy of Science, proposing their own view, according to which History and Philosophy of Science are and *should be* two distinct and different entities. Thus, the neo-positivist model represents the second phase of the dialectical logic: the anti-thesis.

In the writings of Israel Scheffer, Carl Kordig, and Ronald Giere we can recognise a distinctive pattern of arguments aiming to reject Kuhn's model. Their main concern is that in Kuhn's picture, Philosophy of Science loses its *normative* power, which they claim to be its defining feature. According to Scheffler (1967) and Kordig (1971), Philosophy of Science is intrinsically a *prescriptive* discourse on science, which aims to set *methodological norms* for the scientific inquiry. As much as Philosophy of Science is concerned with the scientific method, there is no interest in the actual practices of scientists but in the *standard of rationality* that makes certain theories *scientific*. For the discussion around the standard of rationality, the temporal development of science is useless, since standards are conceived as *logical* entities that are constitutively ahistorical. Consequently, Carl Kordig (1971, ch. 4) writes that even if all scientists in all ages were fundamentally irrational and broke every normative rule of the scientific methodology, this would not affect the scientific method itself.

To support their argument, both Kordig and Scheffler refer to a distinction that was first drawn by Hans Reichenbach and later on became a renowned model among philosophers of science: the distinction between the *context of discovery* and the *context of justification*.[4] This distinction was originally put forth by Reichenbach as a means to distinguish epistemology from psychology. According to Reichenbach, psychology deals with the *actual* processes of thinking taking place in the mind of scientists at work, inquiring the psychological genesis and conditions of scientific discovery (the 'context of discovery'). Epistemology, on the other hand, does not study a real process, but a *logical* substitute consisting of all the logical steps, which are ideally needed to fully justify a scientific assertion. This logical object was called by Reichenbach 'rational reconstruction', and it is intrinsically both ideal and normative: *ideal* because it does not happen in any specific space and time, and *normative* because it corresponds to the steps that a fully-fledged rationality would take in order to justify any scientific claim. Logical reconstruction constitutes the 'context of justification' of science, and it is the only object of epistemology.[5]

In the neo-positivist reading of Reichenbach, the context of discovery is stretched to include not only the psychology of science but all the conditions involved in the genesis of scientific theories, from social to economic and political context. History of Science is thus pointed out as the *par excellence* discipline dealing with the context of discovery, whereas Philosophy of Science is conceived as the study of the context of justification.

Once the distinction between the two contexts and, hence, between the two disciplines is drawn, the final step of the neo-positivist argumentation is to invoke the so-called Hume's law (no 'ought' from an 'is') to rule their relationship. According to Hume's law, normative claims (ought) and descriptive statements (is) are logically independent, that is to say, the latter does not entail the former. Therefore, we cannot derive from stated facts (descriptions) what we ought to do (prescriptions).[6] In the case of science, the application of Hume's law shows that we cannot infer from the mere description of scientists' behaviour how science ought to be conducted. This means that evidence from the context of discovery (History of Science) cannot affect the context of justification (Philosophy of Science). Thus, the relationship between History and Philosophy of Science is severed from the beginning: the gulf between them is the gap between description and prescription, reality and normativity. A dualist model, where Philosophy of Science and History of Science lie on the opposite side of the ought/is dichotomy, opposes the Kuhnian empiricist model (History of Science provides evidence, Philosophy of Science theory). In this new model, the scope of History of Science is limited to a neutral description, whereas prescription is an exclusive domain of Philosophy of Science.

As previously mentioned, in the dialectics that I am presenting, the neo-positivist model plays the role of the anti-thesis. According to Hegelian logic, the anti-thesis negates the thesis by exposing an essential fault within it. I argue that the neo-positivist argument unveils a *real issue* that we find not only in Kuhn's Philosophy of Science, but also in any other research project that aims to describe a scientific activity in the absence of a normative standpoint.[7] Ronald Giere (1973, p. 290) has clearly highlighted this issue: 'If one grants that epistemology is normative, it follows that one cannot get an epistemology out of the history of science – unless one provides a philosophical account which explains how norms are based on facts'.

Neo-positivist criticism shows how the Kuhnian empiricist model structurally lacks the theoretical machinery that is needed to support a normative dimension in Philosophy of Science *because any attempt to bridge the distance between the descriptive and the prescriptive level will result in a violation of Hume's law*. The production of norms of science (epistemological prescription) from the description of scientists' behaviour (historical reconstructions) is indeed a special case of the derivation of normative statements from facts – that is precisely what Hume's law prevents. On the other hand, neo-positivism provides theoretical room for normativity by discerning the context of discovery from the context of justification, as it acknowledges that there is a difference between what science *is* and what science *ought* to be.

However, the price for normativity in the neo-positivist model is too high. The normative dimension of the context of justification is achieved by completely separating the philosophical discourse from the actual scientific practice. Because of the very distance between the context of discovery and context of justification, the object of Philosophy of Science is downgraded to a mere logical surrogate. Historically, the fate of the neo-positivist project of a 'logic of science' has indeed failed mainly because of this criticism. Abstract models of scientific rationality were proved to be incapable of explaining existing scientific practice, as demonstrated by thorough sociological studies. Moreover, the methodological norms appeared to be too abstract to provide any useful guidance to scientists.[8] In 1984, Dudley Shapere declared the end of the neo-positivist normative programme, and acknowledged that an *a priori*, normative definition of scientific methodology could not be provided.[9]

The negation of the thesis (anti-thesis) is thus negated in turn. Should we then return to the thesis, the Kuhnian model, and give up the very possibility of a *normative* Philosophy of Science? Is Philosophy of Science just a theory accounting for historical facts? If so, iHPS does not have any distinctive features, and it should be considered just as another province within the galaxy of STS. However, in the next paragraphs I demonstrate that this is not the case. Indeed, during the debate of the 1960s and 1970s, we find a third position that can be considered as a Hegelian *synthesis* of the Kuhnian and neo-positivist models: the Popperian model of the relation between History of Science and Philosophy of Science.

4.1.3 Popperian normative historically oriented iHPS and exemplary logic

In order to introduce the Popperian model, it is useful to consider the doubts that it raised in Kuhn. In his talk at the well-known conference *Criticism and the Growth of Knowledge*, Kuhn acknowledged that Popper and his school made large use of historical examples in their writings of Philosophy of Science. However, Kuhn (1970a) was puzzled by the fact that Popper strongly rejected his notion of 'normal science', which Popper (1970) even defined as a 'danger' in his paper 'Normal science and its Dangers'. How could this happen as both contenders relied upon History of Science? Was it only because of a different interpretation of the same empirical evidence? I believe the answer can be found by digging deeper into their theories. In particular, Kuhn and Popper differ in the *status* they assign to the History of Science. As explained before, History of Science is for Kuhn a class of phenomena that the historian arranges in narratives and the philosopher tries to explain via a theory. From this point of view, however, historical facts are to be regarded as mere empirical facts. They lack any intrinsic normative value, they are *neutral* as regards epistemological values. History of Science has no internal axiology for Kuhn. Lakatos (1978, p. 135, note 4) perfectly summed up Kuhn's position by dubbing it 'historiographical positivism', since both Kuhn and the nineteenth-century Positivists share the idea that facts and values belong to different realms. According to them, historical facts are indeed a normative-free set of phenomena.

Significantly, the *neutral view of History of Science* underlies both Kuhn and his neo-positivist critics. The only difference between the two is that the former lacks theoretical means to formulate a normative discourse, whereas the latter has such tools. Nevertheless, both share a *neutral attitude* towards the past of science: the History of Science as such is neither good nor bad. On the other hand, Popper considers at least some episodes in History of Science *intrinsic* examples of good science. The role of the Popperian philosopher of science is to *distil* the scientific methodology from these very cases. Popper does not conceive the method of 'conjectures and refutations', as explained in *Conjectures and Refutations* (Popper 1963), as a prescriptive norm that is detached from scientific practice (the neo-positivist model), but as a distillation of the real methodology implemented by *good* scientists. It is pivotal to highlight the role that the notions of 'good science' and 'good scientist' play in the Popperian view of History of Science. Importantly, these *structurally evaluative* terms define the class of historical events that, according to Popper, should be the target of the philosopher of science.

Unlike the Kuhnian philosopher of science, who is supposed to explain most of the historical facts, the Popperian philosopher of science should derive scientific methodology *only* from those events that are recognized as good science. Popper underlines that good science is not a quantifiable matter: good science may be the minority of actual scientific practices. Although he acknowledges that scientists may spend most of their time in a condition of Kuhnian normal science, he claims that the philosopher of science should be interested *only* in the examples of good scientific practices, no matter how sporadic they might be.[10]

From the Popperian's perspective, the examples of good scientific practice take on the role of *exemplars*. History of Science is not thought of as a set of neutral historical facts, but as a *repertoire* of instances of good – and bad – science. If Kuhn's model for History of Science is political history,[11] Popper's model for History of Science is *sacred* history. In sacred history, certain facts and figures have an intrinsic normative value. In Christian hagiographies, for instance, episodes in the lives of saints (*exempla*) are provided *as such* with axiological import. Christians are supposed to find in these episodes exemplars of how to live in accordance with Christian moral principles. They represent exemplars of a good life.[12] In the same way, for Popper the examples of good science have applied the right scientific methodology that should guide the philosopher of science. Popper elaborates on the ancient Latin motto *historia magistra vitae est* (history is life's teacher): the philosopher of science can learn from the History of Science, as it provides examples of good science that have a normative value. Both from a Kuhnian and a neo-positivist point of view, it is difficult to support an educative role of the History of Science, because of the neutral notion of history they both share.[13]

The exemplary view of History of Science closes the Hegelian triad with a synthesis of both the thesis and the anti-thesis. It shares with the thesis (Kuhn's empiricist model) the idea that History of Science is important for Philosophy of Science, but it refuses the Kuhnian neutral notion of History of Science. From the anti-thesis (the neo-positivist dualistic model), it retains the importance of a normative dimension for Philosophy of Science, but rejects the concept of an ahistorical normative standard. Thus, the exemplary model is a synthesis that unifies (in the Hegelian sense of *aufheben*) the previous positions and integrates History and Philosophy of Science in a new way. This integration can be named a 'normative historically oriented the iHPS'.

However, the Popperian synthesis may be judged unstable and, perhaps even question-begging. From a Kuhnian point of view, the idea that philosophers of science should distil scientific methodology only from a few historical episodes, idiosyncratically chosen as good science, could be rejected as a mere *ad hoc* strategy to avoid empirical testing of philosophical theory. On the other hand, from a neo-positivist point of view, the notion of good science may be challenged because of the lack of bases, in case an independent and ahistorical normative framework is not able to justify it, or can be even considered as a violation of the ought/is divide, since methodological prescriptions would be inferred from mere descriptions of scientific episodes.

Indeed, from a strictly logical point of view, the integration of History and Philosophy of Science in Popperian normative historically oriented iHPS is *circular*. Methodology (read Philosophy of Science) is meant to be distilled *from* examples of good science (exemplars, read History of Science). At the same time, examples of scientific practice are regarded as exemplars *because of* their conformity to methodology. This seems to be a logical loop, where the premises of the argument are based on the conclusions of the argument itself – a kind of logical fallacy that is termed *circulus in probando* or *petitio principii*.

However, the Popperian normative historically oriented iHPS remains an innovative approach to the integration of History and Philosophy of Science. Even if its results are inconsistent from a strictly logical point of view, it is worth considering because its core structure can be extensively found today in a specific discourse concerning science: the discourse of *Science Policy*. The following section explains in detail how the circular logic of normative historical oriented iHPS is *real* (in the Hegelian sense of *wirklich*, 'active', 'actual') in the domain of Science Policy. In doing this, I show that Science Policy represents an area where the integration of History of Science and Philosophy of Science has been achieved in practice.

4.2 The integration of History and Philosophy of Science in science policy discourse

This section argues that the exemplary logic, which is the circular integration between History and Philosophy of Science that results in a normative historically oriented iHPS, is a fundamental component of the *Science Policy discourse.* Science Policy consists of diverse topics concerning the allocation of resources for all the different activities related to science. For instance, topics include funding of science, careers of scientists, intellectual property policy and translation of scientific discoveries into technological innovation.[14] Science Policy encompasses all the stakeholders that are involved in the management and governance of the scientific enterprise, from universities to institutions and social actors. Vis-à-vis the academia-based disciplines of History and Philosophy of Science, Science Policy does not take place only at the university. In fact, Science Policy discourse is spread across a large web of actors contributing to proposing, implementing, modifying, and even opposing science policies. University-based discourse on Science Policy is produced under the label of Science Policy and Innovation Studies (SPIS), and it is disseminated through scientific journals, the traditional academic format.[15] Different disciplines (e.g. management, economics, sociology) contribute to this field. On the other hand, the institution-based discourse is spread through different formats, including reference documents, white papers, and political speeches.

A detailed analysis of the structure and features of Science Policy discourse is beyond the scope of the present chapter.[16] Instead, I aim to demonstrate how three prominent examples of Science Policy discourse follow the same logic that governs the Popperian integration of History and Philosophy of Science (namely, the exemplary logic). By using those three instances, I will show how the integration between a *sui generis* History of Science and a *sui generis* Philosophy of Science takes place in the same circular manner that characterises Popper's normative historically oriented iHPS.[17]

The three examples are: Vannevar Bush's (1945) famous report *Science, the Endless Frontier*, Henry Etzkowitz' 2008 *The Triple Helix* and the European Commission (2014) 'self-evaluation form' of the European Research Council's (ERC) Starting and Consolidator Grant (a document provided to participants to self-evaluate the projects they submit to ERC for funding). I chose Bush's report because of its historical importance in shaping Science Policy in the USA. Etzkowitz's monograph is considered an example of contemporary academic discourse on Science Policy. Lastly, the European document provides a sample of documents generated by Science Policy institutions.

4.2.1 Vannevar Bush's science, the endless frontier

During the Second World War, Vannevar Bush headed the United States Office of Scientific Research and Development (OSRD), which managed almost all wartime military research. This included the Manhattan Project, which carried out the wartime research and development that produced the first nuclear weapons. In 1945, Bush delivered to President Roosevelt a report, entitled *Science, the Endless Frontier*, which had a profound impact on post-war Science Policy in the USA. In the report, Bush stresses the strategic importance of scientific research for the future of the country after the war. Science and technology are depicted as the key to assure not only national security, but also economic growth and public health. However, he adds, the essential condition to maximise the potential of science was a strong investment of public money into scientific research. Bush's idea is that the market alone could not provide the capital needed for research projects, as research would have been considered an extremely high-risk investment. Therefore, the money needed for scientific research should be supplied by the government. The main proposal advanced in the report is that the huge funding that the US government granted to science during the war should have been maintained also in times of peace, with the creation of a National Science Foundation.

In Bush's argument, the distinction between *basic* and *applied* science is pivotal. Today, these terms are widely used in Science Policy.[18] However, at the time of Bush's proposal, they were a recent acquisition since they were introduced in the 1920s during the British discussions on Science Policy. Bush actively contributed to shaping their meaning. In *Science, the Endless Frontier*, basic science is famously described as research 'performed without thought of practical ends', resulting in 'general knowledge and an understanding of nature and its laws' (Bush 1945, p. 18). On the other hand, applied research is characterised by research devoted to solving practical problems, *building from the results carried out in basic research*:

> This general knowledge provides the means of answering a large number of important practical problems, though it may not give a complete specific answer to any one of them. The function of applied research is to provide such complete answers.
>
> (Bush 1945, p. 18)

The idea that knowledge flows *unidirectionally* from basic to applied research is central in Bush's argument because it justifies the need of mainly funding basic research. This concept is in contrast to the pre-war US Science Policy, which basically only funded applied research (especially in agriculture and farming).[19]

The strategy used by Bush to justify his proposal builds on certain technological advancements that, he claims, could have been achieved only by progress in basic research. Radar was Bush's favourite example. Radar was a military device that was essential to defeat Germany, and it was developed thanks to basic research in physics. Another example is antibiotics (such as penicillin), which were the outcome of basic bio-medical research and were crucial in saving the lives of thousands of American soldiers during the war.

We can say that Bush's argument combines a *normative proposal* (the Science Policy principle of funding primarily basic research) with a *set of examples* that are patently desirable scientific outcomes, that is to say, they play an *exemplary* role. Thus, Bush's examples have the same purpose as the examples of good science in Popper's normative historically oriented iHPS. Therefore, both Popper and Bush base their arguments on the same logic, the exemplary logic.

In the case of Popper, this logic relates examples of good science (History of Science) to good scientific methodology (Philosophy of Science). In the case of Bush, it relates desirable scientific outcomes to good Science Policy principles. Once again, a descriptive part (the History of Science part) is circularly connected with a prescriptive part (the Philosophy of Science part). Indeed, some episodes, such as Einstein's relativity theory for Popper, and radar and antibiotics for Bush, are described as desirable scientific outcomes, that is to say exemplars. Starting from these exemplars, the authors build on strategies to prepare the ground for more exemplars: Popper formulates the idea of the conjectures and refutations method, while Bush states the principle of funding basic research. The whole process is indeed circular. Bush's manifesto is the first example of circular logic, and it is a model of normative historically oriented iHPS. However, *Science, the Endless Frontier* was published in 1945 and is therefore a historic work. Hence, in the next section, I examine a recent Science Policy document to verify whether the same logic does still exist in current Science Policy.

4.2.2 Henry Etzkowitz's triple helix model of the relationship between university, industry and government

The second instance I consider is Henry Etzkowitz' 2008 monograph *The Triple Helix*. According to Etzkowitz, a closer integration of university, industry and government (the 'triple helix' mentioned in the title) is the key to increase growth and promote innovation in knowledge-based economies. The author proposes several policies to foster this integration. For instance, universities are encouraged to become entrepreneurial by taking on some roles that are traditionally attributed to industry, such as the development of new firms and the capitalisation of knowledge. At the same time, companies are encouraged to develop high-level training programmes and share knowledge by establishing joint ventures. Thus, private companies are supposed to become a form of para-university. Finally, the government is meant to combine traditional regulatory activities with public venture actions by providing public capital to high-risk (but potentially high-gain) research-based companies. Etzkowitz (2008, p. 1) thinks that the university should be the leading force of the triple helix, and he regards it as the 'source of entrepreneurship and technology as well as critical inquiry'.

Etzkowitz's programmatic book is permeated by exemplary logic: as in the case of Bush, the proposed policies are always coupled with instances that exemplify those policies. For instance, the pillars of the entrepreneurial university are the following (Etzkowitz 2008, p. 27):

1   Academic leadership able to formulate and implement a strategic vision.
2   Legal control over academic resources, including physical property, such as university buildings, and intellectual property emanating from research.

3   Organisational capacity to transfer technology through patenting, licensing and establishment of business incubators.
4   An entrepreneurial ethos among administrators, faculty members, and students.

Each of these pillars is explained in Etzkowitz's work as providing tangible examples of universities around the world that have adopted the suggested policies and have thrived thanks to them. For instance, the Renssellaer Poly-technic Institute (Troy, New York) is presented as the first institution that created an incubator (an organisation assisting university spin-offs that have their roots in academic research). Cases presented from Brazil and Africa exemplify the benefits of converting traditional universities into entrepre-neurial universities: according to the author, this transformation has helped to spread an entrepreneurial ethos within Brazilian society, and has solved technological crises in Africa. In Chapter One ('Pathways to the triple helix'), the Boston Area and Silicon Valley in Northern California are pointed out as the best-case scenarios of the application of the triple helix model.

Etzkowitz employs throughout the text the very same strategy: while the best-case scenarios are the evidence of the good policies he suggests, *at the same time*, he considers those policies good because they are required for the very existence of the best-case scenarios. As we can see, Etzkowitz' reasoning shows circularity, a characteristic feature of the exemplary logic. First, certain episodes are chosen to serve as examples of best science, that is exemplars. Second, a *sui generis* Philosophy of Science is extracted from those exemplars, which leads to the creation of a model, namely the triple helix model. The model is then used to explain the necessary conditions that set the stage for exemplary episodes. Finally, this Philosophy of Science is translated into guidelines for Science Policy, guaranteeing that its implementation will lead to new best-practice models.[20] Therefore, we can conclude that the exemplary logic is still present in contemporary Science Policy texts, and, specifically, in the academic context. In order to address the question whether the exemplary logic exists also in the institutional discourse, I will now focus on European Science Policy documents, the final case-study of this chapter.

4.2.3 European Research Council Science Policy

The final instance of exemplary logic in Science Policy I consider here is the current European Research Council (ERC) policy, as it is represented by the self-evaluation form, which is provided by the European Commission (EC) to researchers submitting project proposals to the ERC (EC 2014). Although quite short, this document is particularly significant, since it summarises what kind of scientific research the EC plans to fund, and it illustrates the scientific desiderata of the EC. As I will show, these desiderata can be considered as the implicit normative Philosophy of Science that guides the European Research Council.

First, the research project should be 'ground-breaking', which means that the research should not follow existing research lines but it should be a true step beyond the state of the art.[21] Research funded by the ERC should have the ambition to potentially revolutionise the whole research field by suggesting new paradigms and implementing novel methodologies. Interdisciplinarity is an important asset for projects: interdisciplinary research is more likely to lead to breakthroughs than research that only investigates a specific discipline. Moreover, the project should address important contemporary challenges: it should be 'timely', to put it in the Commission's terms. These pending issues can be either science- or society-related. Scientific challenges include topics that have greatly attracted the attention of the scientific community in the past few years, whereas the societal challenges consist of all those issues that directly affect the life of European citizens. The rationale behind these criteria is that research should have an impact on citizens, as research should be able to improve their lives. In the ERC documents, the flag term that summarises all the features mentioned above is 'frontier research', where the adjective 'frontier' means: frontier of knowledge ('ground-breaking'), frontier among disciplines ('interdisciplinarity'), and frontier between science and society, science and technology, science and industry ('impact'). Research funded by the ERC should be 'on the frontier' by any means.[22]

However, all these notions remain quite abstract, and applying them to real research projects might be challenging. Here is where the exemplary logic comes into play. The ERC and university grant offices supply the participants with several *examples* of previously funded projects. By doing this, they aim to translate the abstract desiderata into something more concrete by providing tangible examples.[23] These examples represent models of good science to which applicants can refer as a guidance to write their own research proposals. These exemplars stand for good research models, which embody the scientific desiderata stated in the European policy. The policy, in turn, is supposed to guarantee that new good models will be produced. The circularity of the argument is, once again, analogous to the previous two Science Policy examples I examined. In the case of Bush, radar and antibiotics were described as exemplary scientific outcomes. These episodes were then used to put forward a funding policy, where the priority was given to basic research instead of applied research. This policy, which can be seen as a normative Philosophy of Science, was in turn justified as the necessary condition to achieve new exemplary scientific outcomes, closing the circle of the exemplary logic. In the case of Etzkowitz, the descriptive and the normative are even more intertwined. The 'triple helix model' is presented as a *description* of the best model of university-government-industry interaction, and at the same time as a *normative* proposal for successful Science Policy. Therefore, we can see that all the three cases I examined are shaped by the *exemplary logic*, which is the same logic at the core of normative historically oriented iHPS. Figure 4.1 schematically depicts the exemplary logic to clarify and better understand its circularity.

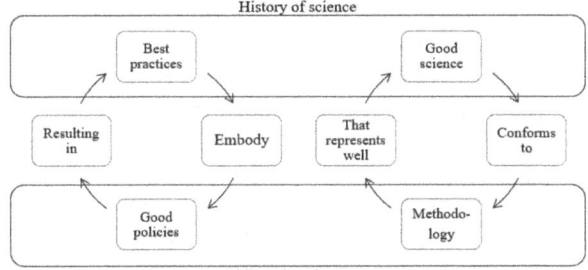

Figure 4.1 Exemplary logic diagram: The exemplary logic in Science Policy discourse
(left) and in Popperian normative historically oriented iHPS (right). The
typical circularity of this logic is clear in the diagram. The upper area of
the diagram concerns the History of Science, while the bottom part the
Philosophy of Science.

4.3 The agenda of Philosophy of Science Policy

The first section of this chapter introduced the distinctive integration between
History and Philosophy of Science that underlies Popper's iHPS, and that I
previously called normative historically oriented iHPS. By examining three
case studies, the previous section showed how the same integration underlies
Science Policy, and demonstrated that this integration is achieved within a
context, which is normative by definition (the Science Policy discourse). Thus,
I concluded that Popper's iHPS and Science Policy are isomorphic if we consider the way they integrate History and Philosophy of Science (the exemplary
logic). This isomorphism strongly suggests the possibility of introducing a
normative Philosophy of Science today, as long as, first, it integrates History
and Philosophy of Science by following the exemplary logic, and, second, it
takes into account the Science Policy context. Therefore, I argue that this
revised normative Philosophy of Science should take the form of a *Philosophy
of Science Policy* (PSP), a term that should maintain the double meaning of
the genitive. In the objective sense, it should address topics in Science Policy,
thanks to its double normative and historical nature. In the subjective sense,
Science Policy should be a favourable context to its growth. The aim of this
final section is to develop a research agenda for a Philosophy of Science
Policy which will be linked to other lines of research that are currently pursued in Philosophy of Science and have been already extensively construed.

Drawing inspiration from Reichenbach's *Experience and Prediction*, I
divide the aims of a Philosophy of Science Policy into a *descriptive* task, a
critical task and an *advisory* task (Reichenbach 1938, pp. 3–16). The
descriptive task pertains to the History of Science itself. On the other
hand, the critical task mainly refers to the Philosophy of Science, while the
advisory task is the context where the integration driven by the exemplary
logic is fully realised.

When pursuing the descriptive task, the philosopher of Science Policy focuses on the categories that Science Policy uses to classify scientific activity. As previously explained, the Science Policy discourse classifies research, by coining new categories to endorse the directives of the policy (for instance 'basic research', 'applied research', 'frontier research', etc.). The philosopher of Science Policy should clarify the different taxonomies of Science Policy and reconstruct how they have emerged in history. Useful contributions to this kind of research can be found in historical semantics, as shown by the work of Désirée Schauz (2014; Schauz and Godin 2016), who tracked down the evolution of the meaning of different terms, for example 'basic research'. Another important contribution is the work of Marc Pauly (2016), who designed an ontological framework to reconstruct the general 'ontology', or basic categories, of policies. In the descriptive task, it is important to take into account the political influence of the choices that were taken while drafting a Science Policy. These choices were the outcome of a complex negotiation between political and epistemological values. Reconstructing the implicit epistemology behind Science Policy is of crucial importance, as it allows us to pinpoint the epistemological values it promotes. In European Science Policy, for instance, the ERC positively evaluated interdisciplinarity, and the analysis of this policy has revealed the epistemological commitment of the ERC towards a specific epistemological value, interdisciplinarity.

The descriptive task lays the foundations for the *critical* task. Philosophers of Science Policy should use models of scientific inquiry developed in Philosophy of Science (such as Kuhn's theory of scientific change, Lakatos' methodology of scientific research programmes, Feyerabend's epistemological anarchism) to *evaluate* and *assess* Science Policy. The ERC's requirement that research should be ground-breaking shall serve as an example of this approach: from a Kuhnian perspective, being ground-breaking is a feature of revolutionary science. However, in the Kuhnian model, revolutionary science occurs during a paradigm crisis. A paradigm crisis happens only when the paradigm has been fully developed, and specifically when scientists identify predictions in the paradigm that are not verified by observations or experiments (these flaws are called by Kuhn 'anomalies' of the paradigm). Therefore, revolutionary science is the final stage of a long process of development of the paradigm. According to Kuhn's (1970b, ch. 9) model, revolutionary and ground-breaking research cannot be generated at will. Hence, from this point of view, the ERC's push towards ground-breaking research may seem like an artificial intervention into the scientific process, doomed to failure if the paradigm has not fully developed. The critical task of PSP consists in flagging up to the ERC this kind of potential problem of the policies, in order to overcome them. This critical task finds a natural ally in Steve Fuller's (1988, 2015) social epistemology, that is defined as the study of what kinds of knowledge are desirable and how they can be best produced, assessed, and disseminated.[24]

Finally, the *advisory* task of PSP is the synthesis between the descriptive and the critical task. The philosopher of Science Policy should step into the arena of Science Policy, advancing proposals for new science policies. This normative mission is the context where the integration between History and Philosophy of Science should be more prominent. While the exemplary concept of History of Science provides theoretical foundations for the normative use of case-studies in History of Science, the Philosophy of Science is the tool to actively design Science Policy. In contemporary Philosophy of Science, economic epistemology, and especially the recent research inspired by Philip Kitcher's (1990) studies on the division of cognitive labour, constitute a research programme which is more closely in line with the outlined advisory task. Recently, some philosophers of science, such as Shahar Avin, have tried to deduce Science Policy consequences from economic epistemology, and, for instance, have proposed to assign a certain amount of research funding via a lottery. Indeed, the mathematical models used in economic epistemology seem to suggest that research would progress more efficiently in this way.[25]

Conclusion

In conclusion, I have introduced an effective and realistic approach to integrate History and Philosophy of Science, which I named Philosophy of Science Policy (PSP). I believe that PSP has a twofold aim. On the one hand, it is an answer to the long-standing debate about the integration between History and Philosophy of Science. To support this idea, I explained how PSP could integrate the two disciplines, taking inspiration from the normative historically oriented iHPS developed by Popper. On the other hand, I argue that it is indeed possible to introduce History and Philosophy of Science into the Science Policy arena, where important decisions about contemporary science and its future directions are actually taken. Indeed, I showed that the exemplary logic, typical of normative historically oriented iHPS, is also the principle of prominent examples of Science Policy discourses (Bush, Etzkowitz, and ERC). Thus, I demonstrated that Science Policy is a fertile ground for the development of the normative and historically oriented iHPS, as long as it takes the form of a PSP. I also outlined a research programme in PSP, where I identified three different tasks: a descriptive, a critical, and an advisory task. To put this research programme into action, it is important to analyse more examples of Science Policy discourse. In this way, the descriptive task will have a wider and more complete coverage. With regard to the critical task, the next crucial step is to clarify and further develop the Science Policy implications of general theories about scientific change, such as the ones proposed by Imre Lakatos and Paul Feyerabend. Finally, as regards the advisory task, the most important step forward is to give voice to philosophers and historians of science in Science Policy circles, where policies are planned and decisions about the future of science are taken. As I stated at the beginning of this chapter, I believe that PSP has to be at the same time a research enterprise and a call for action.

Notes

1 Since the focus is the inner logic of the debate, the presentation will not follow closely the chronology (even though it will not deviate considerably from the real course of events). For a similar reconstruction (but lacking the Hegelian element) see Steve Fuller (1993). See also Thomas Nickles (1995).

2 A thesis that was contested famously by Imre Lakatos (1978) in 'History of Science and its Rational Reconstruction', where Lakatos argues that historical facts about science development are not independent from philosophical theory, but are framed by the underlying philosophy of science of the historian reconstructing them.

3 For the classic criticisms of *The Structure of Scientific Revolutions* by Popper, Lakatos, Feyerabend, Watkins etc., the *locus classicus* is Lakatos and Musgrave (1970). Paul Hoyningen-Huene (2000); Wes Sharrock and Rupert Read (2002); and Alexander Bird (2000) present detailed and insightful reconstruction both of Kuhn's thought and his major critics.

4 A similar distinction is stated by Popper (1959), ch. 1.

5 See Hans Reichenbach (1938), pp. 3–16.

6 See Georg Spielthenner (2017) for a clear and up-to-date introduction to Hume's law in practical ethics.

7 Mainstream STS comes to mind, as well as Lorraine Daston and Peter Galison's 'historical epistemology' pursued in studies such as *Objectivity* (Daston and Galison, 2007), where the authors explicitly claim that their narrative has no normative import.

8 See the discussion of epistemological values developed by Kuhn (1977).

9 Shapere reached the conclusion that science is whatever scientists say it is. See Dudley Shapere (1984), ch. 10 and Steve Fuller (1993).

10 See Popper (1970).

11 In ch. 9 of *Structure*, Kuhn discusses the parallelism between *political* and *scientific* revolutions. He argues that both are inaugurated by a growing sense of crisis, and both aim at changing existing institutions or paradigms in a way which is prohibited by those institutions and paradigms. See Kuhn 1970b, pp. 92–94.

12 See Yoram Hazony (2012).

13 This is true of neo-positivist philosophy of the 1960s. Pre-war logical empiricism exponents, such as Reichenbach and especially Otto Neurath, are closer to the exemplary view of History of Science.

14 Sometimes, also research ethics is counted among Science Policy topics, under the label of Responsible Research and Innovation (RRI).

15 See Ben R. Martin (2012).

16 See Aant Elzinga (2012), for an enlightening historical reconstruction of how Science Policy regimes changed from the Humboldtian university reform to contemporary globalised mega-science.

17 It is worth stressing that I am not arguing that Science Policy amounts to iHPS in practice *simpliciter*. My point is that the discourses by which science policies are justified are shaped by the same logic underlying Popper's integration of History and Philosophy of Science: the circular 'exemplary' logic binding together the normative and the descriptive. Thank you to an anonymous reviewer for highlighting this point.

18 It is worth remembering that the distinction between basic and applied science has been criticised by both historians of science and STS scholars, as unable to capture the reality of scientific practice. See, amongst others, Bruno Latour (1987); Peter Dear (2005); Steven Shapin (2008); and Peter Galison (2008). In the times of Bush, however, these terms were important notions within Science Policy debates. For example, the National Science Foundation (1953) Third Annual Report is entitled, tellingly, 'What is Basic Research?' and, ten years later, the OECD Frascati Manual codified the distinctions between 'basic research', 'applied science' and

'development'. See Désirée Schauz (2014); Benoît Godin (2006); and Elzinga, (2012).
19 See Pietro Greco (2013).
20 Etzkowitz does not himself use the term 'philosophy of science'. However, given the generality of his considerations about science and their normative import, it does not seem wrong to use this term to label some of his ideas.
21 In this analysis, I will focus only on the requirements of the project, leaving aside requirements on the Principal Investigator's CV.
22 The definition of 'frontier research' is available on https://erc.europa.eu/abou t-erc/mission (accessed 11 June 2017). See also the 2005 high-level expert group report *Frontier Research: The European Challenge* (European Commission 2005).
23 They are available on the CORDIS database (http://cordis.europa.eu/projects/hom e_it.html). See also ERC website where funded projects (https://erc.europa.eu/p rojects-figures/erc-funded-projects) and 'ERC Stories' (https://erc.europa.eu/p rojects-figures/stories) are available (all links accessed 27 November 2017).
24 See also Nancy Tuana (2010), which advocates a programme in Philosophy of Science Policy more focused on research ethics and ethical issues in conducting science.
25 Some useful insights may be found in the recent Special Issue of *ROARS Transactions. A Journal on Research Policy and Evaluation*, devoted to 'Research Policy: Insights from Social Epistemology'. See Petrovich and Viola (2018).

Acknowledgements

I would like to thank Luca Guzzardi, Daniele Cassaghi, and Selene Allevi for helpful comments on an earlier draft of this chapter. I would also like to thank all the participants of the reading group 'Politics of Episteme' that took place at the Department of Philosophy of the University of Milan in winter 2016, and Giulia Petrovich for linguistic revision.

Bibliography

Bird, Alexander, *Thomas Kuhn* (Chesham: Acumen, 2000).
Bush, Vannevar, *Science, the Endless Frontier: A Report to the President* (Washington, DC: United States Government Printing Office, 1945).
Daston, Lorraine, and Peter Galison, *Objectivity* (New York: Zone books, 2007).
Dear, Peter, 'What is the History of Science History of? Early Modern Roots of the Ideology of Modern Science', *ISIS*, 96. 3(2005), 390–406.
Elzinga, Aant, 'Features of the Current Science Policy Regime: Viewed in Historical Perspective', *Science and Public Policy*, 39. 4(2012), 416–428.
Etzkowitz, Henry, *The Triple Helix: University-Industry-Government. Innovation in Action* (New York, NY: Routledge, 2008).
European Commission (EC), *Frontier Research: The European Challenge* (Brussels: European Commission Directorate General for Research, 2005).
European Commission (EC), *Self-Evaluation Form: ERC Starting Grant (StG) and Consolidator Grant (CoG)* (2014), http://ec.europa.eu/research/participants/data/ref/ h2020/call_ptef/ef/h2020-call-ef-erc-stg-cog_en.pdf (accessed 27 November 2017).
Feyerabend, Paul K., 'Explanation, Reduction, and Empiricism', in *Scientific Explanation, Space and Time*, H. Feigl and G. Maxwell, eds., (Minneapolis, MN: University of Minnesota Press, 1962), 28–97.

Fuller, Steve, *Social Epistemology* (Bloomington, IN: Indiana University Press, 1988).

Fuller, Steve, *Philosophy of Science and Its Discontents* (London: The Guilford Press, 1993).

Fuller, Steve, *Knowledge: The Philosophical Quest in History* (London: Routledge, 2015).

Galison, Peter, 'Ten Problems in History and Philosophy of Science', *ISIS*, 99. 1(2008), 111–124.

Giere, Ronald, 'History and Philosophy of Science: Intimate Relationship or Marriage of Convenience?', *British Journal for the Philosophy of Science*, 24. 3(1973), 282–297.

Godin, Benoît, 'The Linear Model of Innovation: The Historical Construction of an Analytical Framework', *Science, Technology & Human Values*, 31. 6(2006), 639–667.

Greco, Pietro, 'Introduzione', in *Manifesto per la rinascita di una nazione*, V. Bush (Turin: Bollati Boringhieri, 2013), 7–73.

Hanson, Norwood R., *Patterns of Discovery: An Inquiry into the Conceptual Foundations of Science* (Cambridge: Cambridge University Press, 1965).

Hazony, Yoram, *The Philosophy of Hebrew Scripture* (Cambridge: Cambridge University Press, 2012).

Hesse, Mary, *Forces and Fields: The Concept of Action at a Distance in the History of Physics* (Mineola, NY: Dover Publications, 2005).

Hoyningen-Huene, Paul, *Thomas Kuhn* (Chesham: Acumen, 2000).

Kitcher, Philip, 'The Division of Cognitive Labor', *Journal of Philosophy*, 87. 1(1990), 5–22.

Kordig, Carl R., *The Justification of Scientific Change* (London: Springer, 1971).

Kuhn, Thomas S., 'Reflections on my Critics', in *Criticism and the Growth of Knowledge: Proceedings of the International Colloquium in the Philosophy of Science, London, 1965*, ed. by Imre Lakatos and Alan Musgrave (Cambridge: Cambridge University Press, 1970a), 231–278.

Kuhn, Thomas S., *The Structure of Scientific Revolutions*, 2nd edition (Chicago, IL: University of Chicago Press, 1970b).

Kuhn, Thomas S., 'Objectivity, Value Judgment, and Theory Choice', in *The Essential Tension* (Chicago IL: University of Chicago Press, 1977), 320–339.

Lakatos, Imre, 'The History of Science and its Rational Reconstructions', in *Philosophical Papers*, I ed. by I. Lakatos and A. Musgrave (Cambridge: Cambridge University Press, 1978), 102–138.

Lakatos, Imre, and Alan Musgrave, eds, *Criticism and the Growth of Knowledge: Proceedings of the International Colloquium in the Philosophy of Science, London, 1965* (Cambridge: Cambridge University Press, 1970).

Latour, Bruno, *Science in Action* (Cambridge: Cambridge University Press, 1987).

Martin, Ben R., 'The Evolution of Science Policy and Innovation Studies', *Research Policy*, 41. 7(2012), 1219–1239.

National Science Foundation, *What is Basic Science?* (1953), https://www.nsf.gov/pubs/1953/annualreports/ar_1953_sec6.pdf (accessed 27 November 2017).

Nickles, Thomas, 'Philosophy of Science and History of Science', *Osiris*, 10. 2 (1995), 138–163.

Pauly, Marc, 'A Framework for Ontological Policy Reconstruction: Academic Knowledge Transfer in the Netherlands as a Case Study', *Journal of Social Ontology*, 2. 2(2016), 303–323.

Petrovich, Eugenio, and Viola, Marco, eds., *ROARS Transactions: A Journal on Research Policy and Evaluation*, 6. 1(2018), online available at https://riviste.unimi.it/index.php/roars.

Popper, Karl R., *The Logic of Scientific Discovery* (London: Hutchinson, 1959).

Popper, Karl R., *Conjectures and Refutations* (London: Routledge, 1963).

Popper, Karl R., 'Normal Science and its Dangers', in *Criticism and the Growth of Knowledge: Proceedings of the International Colloquium in the Philosophy of Science, London, 1965*, ed. by Imre Lakatos and Alan Musgrave (Cambridge: Cambridge University Press, 1970), 51–58.

Reichenbach, Hans, *Experience and Prediction: An Analysis of the Foundation and the Structure of Knowledge* (Chicago, IL: University of Chicago Press, 1938).

Schauz, Désirée, 'What is Basic Research? Insights from Historical Semantics', *Minerva*, 52(2014), 273–328.

Schauz, Désirée, and Benoît Godin, 'The Changing Identity of Research: A Cultural and Conceptual History', *History of Science*, 54. 3(2016), 276–306.

Scheffler, Israel, *Science and Subjectivity* (Indianapolis, IN: Bobbs-Merrill, 1967).

Shapere, Dudley, *Reason and the Search for Knowledge* (Dordrecht: Kluwer, 1984).

Shapin, Steven, *The Scientific Life: A Moral History of a Late Modern Vocation* (Chicago, IL: Chicago University Press, 2008).

Sharrock, Wes, and Rupert Read, *Kuhn: Philosopher of Scientific Revolutions* (Cambridge: Polity Press, 2002).

Spielthenner, Georg, 'The Is-Ought Problem in Practical Ethics', *HEC Forum*, 29. 4 (2017), 277–292.

Tuana, Nancy, 'Making Philosophy of Science More Socially Relevant', *Synthese*, 177. 3(2010), 471–492.

5 Historical epistemology and the 'marriage' between History and Philosophy of Science

Matteo Vagelli

Introduction

The coming together of history of science and philosophy of science is a far larger and more complex phenomenon than that which, almost exclusively referencing the Anglophone world and the work of Thomas S. Kuhn, goes by the name of the 'historical turn in philosophy of science'.[1] The term 'historical epistemology' instead has come to refer to a wider array of programmes for the combination of history and philosophy of science, ranging from the end of the nineteenth century to the present day.[2] But the heterogeneity of such programmes as well as the relatively recent proliferation of empirical studies grouped under the umbrella of historical epistemology have given rise to a complex and fragmented panorama constituted by what has been seen as a lack of coherence.[3] Indeed, philosophers and historians with very different backgrounds and interests have appealed to 'historical epistemology', especially from the 1990s onwards, and their work has been flanked by questions about the nature, objects and methods implied by historical epistemology.[4] What is remarkable is that the field of historical epistemology, despite its current proliferation, seems permanently haunted by questions relative to its nature, limits and ultimate tasks. Yves Gingras has critically remarked that the current heated discussion about the meaning and use of the expression 'historical epistemology' is a will-o'-the-wisp, a transitory 'brand into the market of ideas'.[5] In his view, 'historical epistemology' is the wrong name for an old programme: the sociology of knowledge, a longstanding historicist programme increasingly taken up since the 1970s by social studies of science proposing what Gingras calls a 'sociological theory of scientific knowledge'.[6]

I believe, in contrast to Gingras, that the discussion of historical epistemology in fact revolves around enduring difficulties in conceptualising the correct or most fruitful interaction between history and philosophy of science. In the next section of this chapter, I trace the recent questioning of historical epistemology back to the enduring and prevalently Anglophone debate over the 'marriage' between history and philosophy of science. I argue in the third section that this renewed interest can in turn be fruitfully situated in a French philosophical context. My main point is that French historical epistemology (*épistémologie historique*) provides an example of a fully integrated approach between history and philosophy of science – one which, if carefully studied, could bring light to current Anglophone debates. In the fourth and last section of this chapter, I illustrate the distance of *épistémologie historique* from the contextualism and historicism, which, according to Gingras, characterise social studies of knowledge.

5.1 The 'marriage' debate: how can History of Science and Philosophy of Science be combined?

The so-called debate over the 'marriage' between history and philosophy of science arose in the 1960s, picked up momentum at the beginning of the 1990s, and, thanks to renewed interest, has continued on into the twenty-first century.[7] The general problem of whether and how philosophy of science and history of science might be combined contains different sub-problems, ranging from the causes of scientific change and the best way to assess its rationality to the theory-ladenness of historical data and the very nature of philosophical analysis itself. In this section, I limit myself to what I take to be the main difficulty at the core of this debate. In a 2011 article, Jutta Schickore argues that the fatal flaw condemning philosophy and history to be endlessly mismatched is the 'confrontation model', in which the two disciplines are not meshed to form a new approach but assembled as pre-given building blocks.[8] In my view, Schickore's assessment illuminates the two main tendencies within this debate: one assigning philosophy a guiding, normative role over history, the other giving predominance to history and fostering a more descriptive, empirical union between the two domains. The combination of history and philosophy of science has mainly taken place in a twofold manner, which, for the sake of brevity I will name respectively the *a priori-normative* and the *empirical* approaches.[9]

In the first case, a normative role is assigned to philosophy, which is understood as an abstract reflection upon the criteria of inference-drawing, theory formation or of rational theory change. Philosophy of science in this sense does not search for an understanding of what science is (even less for what science has been in the past) but for an explanation of what science ought to be in principle.[10] Hence, philosophy is *a priori* normative and therefore aims to develop its own 'theory of theories' regardless of actual scientific theories. This position is held by Norwood Russell Hanson (1962) and Roland Giere (1973) among others. Finding no strong conceptual rationale for history and philosophy of science, Hanson concludes that history of science is 'irrelevant' to philosophy, whereas Giere believes the relationship between the two disciplines to be a mere 'marriage of convenience', rather than an 'intimate relationship'.[11] Nevertheless, a philosopher can 'take a look' outside of philosophy for instructive examples of inferences actually drawn by scientists in order to adjust his or her generalisations.[12] According to this view, therefore, it is not a history of science that is needed by philosophers, but a closer look at the actual practices of contemporary science. Accordingly, Giere believes it was not history that made Norwood R. Hanson's, Thomas Kuhn's or Stephen Toulmin's criticisms of, and alternative proposals to, logical empiricism effective, but rather their appeals to real science. The fact that they referred to the science of 'Kepler and Darwin, rather than R. P. Feynman and J. D. Watson may have been incidental.'[13] In Giere's view, the flaw undermining logical empiricists' attempts at rationally explaining science was self-referentiality. This is why he concludes that a philosophy of science that referred to actual scientific practice would be at best 'convenient', because it would be less self-referential.

In the second approach to history of science, history is taken as the *laboratory* where philosophical claims can be tested. Kuhn has for example suggested that the aim of the history of science is to provide examples and evidence for philosophers' generalizations.[14] More particularly, Kuhn hints at a conception of philosophy of science aimed at providing a *theory of science*, one which must be subjected 'to the same scrutiny regularly applied to theories in other fields' and whose method of data collection is borrowed from the sciences themselves.[15] As a consequence, Kuhn could not propose any fully integrative approach between history and philosophy of science; on the contrary, he urged 'history and philosophy of science [to] continue as separate disciplines. What is needed is less likely to be produced by marriage than by active discourse'.[16] Such dialogue between the two disciplines, he argued, should be inter-disciplinary rather than intra-disciplinary. The differences between the history and the philosophy of science, Kuhn maintained, should not be subverted, since no one can practise history and philosophy of science at the same time, but only alternately.[17]

A much more prominent advocate of this methodological naturalisation is Larry Laudan, who advanced the idea that the philosopher of science should use historical cases as data to produce an empirical theory of theory choice.[18] Through a collective effort carried out at the Virginia Polytechnic Institute, Laudan promoted a philosophy of science founded upon the scientific practice of constructing and testing empirical theories. In 2011, well into the debate about the nature and meaning of historical epistemology as a new programme for integrating history and philosophy of science, we find a similar proposal by Philip Kitcher. In line with his advocacy for methodological naturalism, Kitcher deployed the metaphor of the laboratory of the history of science to emphasise the need for philosophy to avoid becoming mere armchair reflection.[19]

Over the years it is the empirical stance that has seemed to prevail, leading to a naturalisation not only of methods but also of its very object of inquiry. Giere, for instance, has recently revised his position by abandoning normativity and adopting a fully naturalised approach. He has come to conceive of philosophy as an empirical and fully naturalised enterprise, advancing theories about science that are liable to be true or false depending on whether they are in accordance or discordance with historical records.[20] In particular, Giere claims philosophy should be naturalised: reduced on the one hand to the cognitive sciences and to the sociology of science on the other hand.[21] It should be remarked, however, that in both the *a priori* normative and the empirical, descriptive approaches the aim of philosophy is to produce a '*theory* of science', a '*theory* of scientific theories' or a '*theory* of conceptual change'.

One cannot fail to notice that the contemporary use of the term 'historical epistemology' has been accompanied by the same kind of questions that characterised the 'marriage' debate. Lorenz Krüger, the principle inspiration for the creation of the Max Planck Institute in Berlin, was deeply involved in the marriage debate, at least from the end of the 1970s on. In his methodological writings, he fostered a hermeneutic-historicist approach whose main tenet was the idea that the relationship between history and philosophy of science was a 'marriage for the sake of reason'.[22] This continuity between problems and the terms deployed to articulate them is what makes it possible for Gingras to claim that 'recent discussions of the term "historical epistemology"' provide us with 'an interesting example of branding in the field of (Anglo-Saxon) history and philosophy of science'.[23] Feest and Sturm, puzzled by the term 'historical epistemology', moreover have raised the following questions: 'is history necessary for epistemology? Is it useful? If so, in what ways and with what consequences? … How should the relation between philosophy of science and history of science be understood? Is it an intimate relationship, or a marriage of either convenience or reason?'.[24] The persistence of the terms of the 'marriage' debate in the discussion of historical epistemology means that, despite the 'historical drive' of the 1960s, there is still need for integration between history and philosophy of science.

5.2 Épistémologie historique: an *a posteriori* normative approach to the History of Science

It is surprising that contemporary debates about historical epistemology rarely refer to the French tradition of *épistémologie historique* and that, when such references are made, they are brief and unquestioning. In this section, I would like to begin to remedy this neglect by showing, on the one hand, how a comparative study (in this case between the French and Anglophone discussions) is a useful move toward the integration of history and philosophy of science. On the other hand, I will show that the historicisation of epistemology was not a given in the French philosophical context, where the problem of how to combine epistemology and history of science was seriously taken up and actively discussed.

In shifting our attention to the French tradition of historical epistemology, it is important to make a preliminary point concerning the aforementioned 'theoretical attitude' of philosophy of science. The term '*épistémologie historique*' has a natural sense in French, and its English version, 'historical epistemology', does not resonate to the Anglo-Saxon world. This is because the English word 'epistemology' indexes what French philosophy refers to as theory of knowledge, or gnoseology. The French word '*épistémologie*' is instead closer to '*philosophie des sciences'*, the critical study of the principles, hypotheses and results of the diverse sciences, which is interested in determining their logical (and not psychological) origin.[25] Since Auguste Comte, this study has been conceived in France as *necessarily* or *intrinsically* historical, rather than theoretical or methodological.[26]

Qualifying epistemology with 'historical' would therefore be almost redundant within a French philosophical context. Nevertheless, this does not mean, as commentators often surmise, that the implications of the different ways to historicise epistemology were not an issue in this context.[27] On the contrary, as Cristina Chimisso has shown, the question of whether and how to combine philosophy and history – at first meant as the history of philosophy, then also as the history of science and general history – was a major preoccupation for a generation of French intellectuals who gained academic recognition at the turn of the twentieth century.[28] In this period, a naturalising trend with two poles can be detected among the widespread efforts to combine history and philosophy: one side aiming to frame philosophy through all-embracing historical syntheses, the other trying to neutrally draw on historical records to sustain philosophical claims. The former position was predominantly promoted by Henri Berr and his *Centre de synthèse historique*, whereas the latter was held by Émile Bréhier and Leon Brunschvicg. On the one hand, Berr's idea was that every particular history was part of a 'total history', or rather, an historical narrative that would not exclude any kind of material or disciplinary approach. Berr aimed to extend the methods of the natural sciences to the human sciences, thus making history a scientific and objective discipline. In particular, Berr favoured an inductive method that could generalise the results of large collections of data, like bibliographic entries.[29] If in Bréhier's view, on the other hand, history is to the philosopher what the experimental method is to the scientist, Brunschvicg's *a posteriori* study of the mind likewise needed the laboratory of 'history' to observe reason at work.[30]

Chimisso rightly presents Gaston Bachelard's and Georges Canguilhem's *épistémologies* as the height of such attempts to combine history and philosophy. In contrast with the empirical and naturalising trend just described, Bachelard and Canguilhem, while retaining an *a posteriori* approach, produced a *normative turn* in historiography that fundamentally opposed both the ideal of a total history and that of an objective, neutral observation and collection of historical facts and data.[31] In what follows, I will concentrate mainly on Canguilhem, whose historiographical reflection is particularly fitting to the 'marriage' debate – a fact which makes the failed dialogue between the Anglo-American and the Continental traditions all the more unfortunate.

Georges Canguilhem (1904–1995) is a key figure of French *épistémologie historique* – not only was he Michel Foucault's *maître*, but he himself was also the 'heir' of Gaston Bachelard: he took from Bachelard the Chair of History and Philosophy of Science at the Sorbonne and the directorship of the *Institut d'histoire et de philosophie des sciences et des techniques* (IHPST). In his 1966 talk titled 'The object of the history of sciences', presented at a conference in Montreal, Canguilhem challenged the historiography of science. He argued that by reassessing the relation of philosophy to science, the philosopher could attain privileged access to the history of science, more so than scientists and historians themselves. Canguilhem's paper aimed to better understand the nature and targets of history of science through a sharper understanding of its object. In order to make this point, Canguilhem identified a 'properly *philosophical* reason to engage the history of science:

> The strictly philosophical reason [to do the history of science] comes from this: without reference to an epistemology, a theory of knowledge would be a meditation on the void, and without relation to a history of sciences an epistemology would be a less important labor which was completely superfluous to the science of which it pretends to speak.[32]

On the one hand, without reference to the history of science, philosophy would be a mere repetition of the sciences, but, on the other hand, without reference to epistemology, history of science would be reduced to mere chronicle. Epistemology and the history of science from this perspective must collaborate and integrate one another's perspectives: epistemology reclaims history in order to sort out the dialectic of conceptual rectification that constitutes science; history of science borrows from epistemology the knowledge-values currently ordering the chronological succession of theories according to intellectual growth.[33] Rather than independent and endlessly recombining, from this perspective history and philosophy of science are internally related. This passage by Canguilhem should therefore be read as the refusal of any confrontational model between history and philosophy of science.

The core argument of Canguilhem's text concerns the very object of history of science. He argues that many of the questions that can be raised with respect to the nature and function of history of science (*who* does the history of science? *where*, in which institutions? *why*, for what reasons?) are all fundamentally linked to this: *what* is history of science the history of? *What* is history of science *about*? It is because this simple question ('what is the object of the history of science?') has been evaded that the philosopher's privileged access to the history of science gets overlooked. The reason that such a question has been evaded or unasked may seem obvious: the object of the history of science is science, just like the object of any science, say crystallography, is crystals, a particular object of inquiry. Yet, Canguilhem suggests that, in the expression 'science *of* crystals', the preposition 'of' does not play the role of a genitive expressing possession (like being the owner *of* a dog). Rather, according to Canguilhem, it indicates science or knowledge *on, upon, about*, crystals.[34] We should therefore ask what does it mean to have knowledge *upon* science? Are we supposed to build knowledge *upon* science in the same way we build knowledge *upon* crystals? A discussion of the *kinds* of objects at play is therefore in order.

In this context, Canguilhem distinguishes three types of objects: natural, scientific, and historical objects. The object of history of science belongs to the last category: it is an object that is intrinsically historical, whereas science is knowledge of a natural object – one that has no history. Crystals, the object of crystallography, might have a history only in the sense that the earth and its minerals have a history, but this history itself is something already there, pre-given to the historian of science.[35] Alternatively, the object of history of science is permanently unachieved, always in the making. Natural objects (e.g. crystals) do not correspond to scientific objects (e.g. to crystals defined in relation to the constancy of facet angles and the regularity of truncation according to systems of symmetries), as these are constituted by science itself and determined by scientific methods. Scientific objects can be said to be second to natural ones, but they do not derive from them. The same goes for the object of the history of science, which is second to the scientific object but is not derived from it. History of science constitutes and carves out its own object, which is, as Canguilhem says, the 'historicity of scientific discourse' itself, in so much as this historicity 'represents the carrying out of an internally law-governed project, but one which is traversed by accidents, retarded or deflected by obstacles, interrupted by crises, i.e. moments of judgment and of truth'. The history of crystallography is therefore a discourse on the dynamic of historical change of discourses on the nature of crystals.[36]

Canguilhem is here reminding us that, if we do not bear these distinctions in mind, there is a constant risk of conflating the object of history of science and the scientific object. This reminder has important consequences for how we conceive the relation between history and philosophy of science. According to Canguilhem there are two fundamentally distinct ways of carrying out the connection between the history and philosophy of science, i.e. two distinct ways in which epistemology and history of science can relate to one another (and this takes us back to the 'marriage' debate): history can be thought of as the *laboratory* of philosophy, or philosophy (and epistemology in particular) can be history's *tribunal*. Canguilhem argued that it is the laboratory-theory, usually associated with an 'experimental theory of the human mind', that has gained the favour of the majority of specialists.[37] But Canguilhem warned that such a conception 'turns back to copy the relation between the history of sciences and the sciences of which it is the history from the relation between that science and the objects of which it is the science'.[38] Extending the experimental relation from science to its history reveals a rather narrow understanding of the functioning of science, since the experimental relation is only one of the possible ways in which science relates to its objects. Behind the extension of the laboratory model from science to the history of science there is the enduring idea that science is reducible to a unique and eternal scientific method 'slumbering at times, vigilant and active in others'.[39] The image of the history of science as a 'mental microscope' that enlarges pre-existing objects is unable to get at the historicity of science, since it only presupposes the 'injection of duration into the exposition of scientific results'.[40] This attitude is only deceptively historical and in fact prevents any access to the historical dimension of science. This is because, as Canguilhem wrote in 1963, in a text dedicated to the role played by history of science in Bachelard's epistemological works, a 'mental microscope does not distinguish between a difficulty and an obstacle, between delay and wander ... A microscope does not judge. A microscope might detect a movement, but it cannot reveal a dialectic'.[41] That is why, in his 1966 talk, Canguilhem argues that it is not a laboratory but a tribunal that is needed to capture and render visible the dialectic of conceptual rectification constituting science:

> In order to understand the function and significance of a history of sciences one can oppose to the model of the laboratory that of the school, or of the tribunal, of an institution and of a place where judgments are brought to bear on the past of knowledge, on the knowledge of the past. But here one needs a judge. It is the epistemologist who is called to furnish history with the principle of judgment by teaching it the most recent language spoken by some science, chemistry for example, and in thus permitting it to retreat into the past, back to the time when this language ceases to be intelligible or translatable into any more loose or more commonplace language which was spoken before.[42]

Canguilhem is here arguing that if one conflates the scientific object with the historical object, one is led to use the metaphor of the laboratory, and hence to construct history of science as itself a science. In order to do away with this wrongheaded association, one has to maintain the specificity of the object of the history of science through the adoption of the alternative historiographical model, that of the tribunal.

Both the metaphor of the tribunal and of the school can be traced back to Bachelard, who taught secondary school classes (*lycée*) before taking his position at the Sorbonne. Bachelard claimed that philosophers should instruct themselves at the school of the sciences and produce a history of science which follows accordingly.[43] In continuity with Bachelard, Canguilhem held that epistemology helps discriminate between lapsed and sanctioned history (*histoire périmée ou sanctionnée*), or rather, between those theories that are recognised as being still part of an actual science and those that are relegated to the repertoire of imaginary, abandoned beliefs.[44] But Canguilhem also advanced an important additional warning concerning the meaning and function of the judgments the epistemologist is demanded to produce:

> A judgment on this matter is neither a purge nor an execution. The history of sciences is not the progress of sciences in reverse, i.e. the putting into perspective of outmoded stages whose truth is today on the point of disappearing. It is an effort to enquire into and give an understanding of the extent to which outmoded notions or attitudes or methods were, in their time, successful; and consequently of the respect in which the outmoded past remains the past of an activity for which it is necessary to retain the term 'scientific'.[45]

Here Canguilhem means that history of science is not written once and for all, but is, on the contrary, always unstable, always in need of rectification. As a result, our scientific past can not only change, but in fact necessarily and continuously changes through the advancement of both scientific production and historiographical reflection.

Before coming back to the 'marriage' debate, I should dwell for a moment on the consequences that this normative stance had on the histories of science that Canguilhem produced. While Bachelard dealt mostly with geometry, physics and chemistry, Canguilhem focused on the medical and the life sciences.[46] The physiological concept of reflex, for instance, was the object of his secondary doctoral thesis, directed by Bachelard. Standard historiography traces the concept of reflex action back to a mechanistic framework and credits Descartes with its discovery. This conclusion shows, according to Canguilhem, a poor understanding of physiology by historians of science. A neural reflex is currently defined in physiology as a spontaneous movement caused by the peripheral neural system, without the involvement of the central system. This distinctive feature cannot be found anywhere in Descartes' texts, and Canguilhem shows that it in fact belongs to the physician Thomas Willis (*De motu musculari*, 1670) and the vitalist and animist framework inspiring his work.[47] Since Descartes and Willis, this concept has undergone three relevant revisions – in clinical medicine, in physiology and in psychology – and any historical account that does not observe these revisions has not been sufficiently instructed by epistemology. A triumphalist history of physiology ascribes the invention of reflex to the mechanist framework, because mechanistic philosophy was, from the nineteenth century onwards, dominant, while the vitalist one was considered lapsed.[48] In this 'objective' account, the element of judgment is disguised or 'sanitised', so to speak.

However brief, this glimpse at Canguilhem's historiographical production shows that his 'normative turn' consisted in the recursive use of an actual norm or value established in current science on the history of that science itself. This particular type of history thus takes the name 'recurrent history'.[49] Bachelard developed the idea of a recurrent history in order to break with teleological narratives – theories growing spontaneously out of everyday experience and flowing spontaneously into another one. For both Bachelard and Canguilhem, the history of science cannot be 'objective' or 'neutral': *épistémologie* is needed to discriminate between what is *périmé* (lapsed) and what is *sanctionné* (approved) in the history of thought.

5.3 Normative, Whig and presentist history

I now turn to show how this 'French' discussion can provide resources for the Anglophone context of the 'marriage debate'. One of the major terms of this debate, as mentioned above, is its sub-problem of the theory-ladenness of historical data. This preoccupation, which cast a negative light over all 'philosophical' uses of history, emerged in particular among advocates of the empirical approach to bringing together history and philosophy. It concerns the risk of a biased use of history by philosophers seeking raw material as external evidence to prove their theories. This worry is multi-faceted, and it may appear in the literature under the name of 'dilemma of case studies'.[50] Indeed, of the many ways in which philosophy can be seen to be 'altering' history, the risk of anachronism is the most feared – at least since Herbert Butterfield's *The Whig Interpretation of History* (1931). This fear of *biased* history often turns into striving for objectivity. The ideal of a 'sanitised' history is claimed as the only one providing some support to philosophical argumentations. But this ideal, urging philosophers to appeal to history only as an external source to be treated as independent seems to reinforce the 'confrontational model' and hinders real integration between historical and philosophical perspectives.

Hasok Chang has interestingly challenged this view and claimed that, in reality, 'we have never been Whiggish'.[51] Chang distinguishes Whiggism from presentism and triumphalism, three different historiographical attitudes that are usually conflated. He regards the second, presentism, as unavoidable given the fact that the historian has no choice but to be in the present. Whiggism, on the contrary, is a form of presentism underwritten by progressiveness, the idea that the present is superior to the past. Historians' attempts to shun all epistemic judgments in the pursuit of an ideal objective and neutral history is only, according to Chang, 'a judgmental stance disguised as non-judgment'. Indeed, it ends being a form of 'triumphalism', an 'uncritical celebration of anybody who won at the time regardless of whether he was right by today's standards'.[52] For Chang, both internalist historians and sociologists working within the Strong Programme tend to uphold the same ideal of historiographical neutrality. Indeed, the principle of symmetry calls for the value of truth itself to be bracketed, so to speak, so that scientific 'winners' and 'losers' are treated in the same manner.[53]

The normative turn characterising Bachelard's and Canguilhem's approaches could appear to Larry Laudan, but also to Bernard Cohen and many other 'anti-presentist' philosophers and historians of science, as a kind of Whiggish, naively outdated history – a biased reconstruction that relies on current values and organises the past accordingly.[54] As I have just shown, their historiographical approach took the form of a recurrent history in which the epistemologist, through a regulated use of anachronism, imposes on the history of science the norms and values characterising the science contemporary with his or her analysis. But it is in fact this standpoint, I argue, that grants epistemology the role, in Canguilhem's own words, of teaching the historian 'the most recent language spoken by some science'.[55] In this respect, Canguilhem maintained that

> there is a clear difference between retrospective critical evaluation of the scientific past in the light of a present state of knowledge (which is certain, precisely because it is scientific, to be surpassed or rectified in the future) and a systematic, quasi-automatic application to the past of some standard model of scientific theory.[56]

Various attempts at 'naturalising' the history of science, that is, at making history of science as scientific and objective as science itself by using its very methods, would only amount, in Canguilhem's eyes, to slightly different versions of a 'scientific inquisition' of past theories. These histories of science would be 'ideological', in that, as Canguilhem claimed in his 1969 talk titled 'What is a scientific ideology?', history of science can be ideological if it has a false consciousness (*fausse conscience*) of its object: the closer it thinks to be to its object, the more it misses it.[57]

As we have seen, especially in 5.1, the general tendency towards a progressive naturalisation of philosophy of science is propelled by the ambition of 'being closer' to the sciences, both their objects and their methods. In 5.2, we saw how Canguilhem's historical epistemology instead introduces a distance from its object – a distance that is opened by acknowledgement of the 'artificiality' of the object of history of science, which is constructed by the historian. Bachelard and Canguilhem seem therefore to endorse a particular kind of perspectival history, whose distinguishing feature is the establishment of a dynamic relation between the past and the present. In this respect, Canguilhem's and Bachelard's recurrent histories, seen from a larger perspective, instantiate a particular form of presentism, one which avoids triumphalism and above all objective or neutral narratives.[58] As a consequence, what becomes clear is the incompatibility of this historiographical approach with both the principle of context-dependence and that of symmetry between the 'winning' and the 'losing' sides of a scientific dispute, two of the main tenets of both historicism and the sociology of knowledge.

Conclusion

I started with the remark that philosophy was not suddenly historicised at the start of the 1960s. Rather, the oscillation between a normative and a descriptive relation between history and philosophy of science created the situation in which the relationship between the two became unstable. History and philosophy did not merge into a single discipline, like a chemical compound, but remained independent entities that gave birth to a new umbrella term: the hyphenated history-and-philosophy of science. Turning to 'French' epistemology, however, one finds an example of the integration of history-and-philosophy of science in the resolutely normative (but *a posteriori*) approach. Analysis of French historical epistemology enables us to single out three methodological points important for an integrated approach in history and philosophy of science: first, it is not enough for the philosopher of science to get closer to scientific practice but it is the historicity of science itself that he or she should address. As a consequence, the confrontational model dominating the 'marriage' debate is inadequate: history and philosophy of science should be intrinsically integrated with one another from the start. This also means that the laboratory model prevalent in naturalised approaches to history and philosophy of science is wrongheaded: epistemology is normative, and the philosopher plays the role of a courtroom judge with respect to history of science. These insights, formulated by Canguilhem mainly in the 1960s, seem all the more fitting today in a landscape of increasingly naturalised historiographical approaches and an expanding field of digital humanities promoting quantified, data-driven and distant reading methods. This is why integrated history and philosophy of science cannot do without an integrated history of philosophy of science – one which is comparative and takes into account both Anglo-American and so-called Continental histories and philosophies of science. In this sense, historical epistemology, in both its French and its contemporary Anglophone iterations, should be understood as a dynamic conceptual arena for continued discussion of how we might effectively integrate history and philosophy of science.

Notes

1 See Bird 2008.
2 See Rheinberger 2010. In Rheinberger's account, the 'historicizing of epistemology' is a long-running process borrowing from both the positivism of Ernst Mach and the conventionalism of Henri-Poincaré, the phenomenology of Edmund Husserl and Martin Heidegger as well as the historical-philosophical works of Ian Hacking and Bruno Latour, among others.
3 The creation, under the flag of historical epistemology, of the Max Planck Institute for the History of Science (MPI) in 1994 catalysed this proliferation. Some examples of the diversity of themes involved in historical epistemology include analysis of the birth of the 'modern fact' by the sociologist and historian Mary Poovey (1998), the study of the origins of writing by Peter Damerow (2006), and the study of the emergence of sexuality by A. I. Davidson (2001). On the accusation of incoherence among historical epistemologies, see Gingras 2010, pp. 444–447 and Feest and Sturm 2011, p. 286.
4 In this respect, it is sufficient to look at the most recent international conferences on this subject: "What (good) is Historical Epistemology?" organized by the MPI in Berlin in 2008 (Feest and Sturm 2011) or "What is Historical Epistemology?" and "What Does Historical Epistemology Want?", the respective titles of two discussions held at Columbia University in 2008. For a discussion of the meaning of the term "historical epistemology" in relation to the work being done at the MPI, see Hacking (1999; 2002).
5 Gingras 2010, p. 441.
6 Ibid., p. 447.
7 See for instance Domski and Dickinson (2010) and Laudan and Laudan (2016). Schickore (2011) provides a good digest of publications and events concerning the relation between history and philosophy of science.
8 Schickore (2011). I find Schickore's overall reconstruction of the 'marriage' debate compelling but I tend to disagree on her reading of Canguilhem.
9 I am here partially reformulating Burian (2001), who distinguished between a 'top-down' and a 'bottom-up' kind of relation between history and philosophy of science: in the former, philosophical claims are meant to be tested against empirical evidence; in the latter, generalisations are drawn on the basis of the available historical records.
10 This is one formulation, given by R. Giere (1973), of what is known as the 'is-ought problem'.
11 The reference here is to the title of Giere's 1973 paper 'History and Philosophy of Science: Intimate Relationship or Marriage of Convenience?' and to its conclusion:

'the primary relationships for philosophy of science are with philosophy and science. Likewise, the primary relationships for history of science are with history and science. What they have in common is science. But this common interest is not a sufficient basis for other than a marriage of convenience' (Giere 1973, p. 296).

12 Ibid., p. 290.

13 Ibid., p. 290. This position is not dissimilar from that held by Kuhn, who maintained that philosophy of science should be improved and made less abstract by resorting to history of science in order to bridge the gap between the sciences and provide data and examples. But present science for Kuhn is better than that of the past in terms of being a proximate source of first-hand information about scientific practice (see Kuhn 1977, pp. 5, 7).

14 Kuhn 1977, pp. 5, 7.

15 Kuhn 1962, pp. 8–9: 'How could history of science fail to be a source of phenomena to which theories about knowledge may legitimately be asked to apply?'

16 Kuhn 1977, p. 20.

17 Ibid., p. 4. Kuhn used to say that he wore two hats, one as a historian of science and another as a philosopher of science, but, as has been argued by many scholars (see for instance Galison 1981, p. 72), he did not wear them simultaneously.

18 Laudan 1987, p. 24: 'I am suggesting that we conceive rules or maxims as resting on claims about the empirical world [if I do this, I pursue that], claims to be assayed in precisely the same ways in which we test other empirical theories. Methodological rules [...] are a part of empirical knowledge, not something wholly different from it'. Cf. Laudan 1989, p. 11.

19 See §3, 'History as the Methodologist's Laboratory', in Kitcher (2011). The paper by Kitcher is part of Feest and Sturm (2011), which illustrates the continuity between the 'marriage' debate and the debate over the sense and use of the term 'historical epistemology' in the Anglophone world.

20 See Giere in Mauskopf and Schmaltz 2012, p. 61: 'I came to the conclusion that philosophy of science should be transformed into something like the theory of science. That is, philosophers should be in the business of constructing a theoretical account of how science works. Philosophical claims about science would then have the status of empirical theories. In short, the philosophy of science should be naturalized. This means, among other things, giving up pretensions to finding autonomous standards for the practice of science'.

21 Giere 1988.

22 Krüger 1982. See also Krüger 2005[1978]. See Klodian-Coko's contribution to the present volume for an example of the hermeneutic-historicist approach in history and philosophy of science.

23 Gingras 2010, p. 441.

24 Feest and Sturm 2011, p. 287.

25 Lalande 1972, pp. 293–294.

26 This is the sense in which Rheinberger uses the term 'epistemology': to indicate a reflection 'on the historical conditions under which, and the means with which, things are made into objects of knowledge. It focuses thus on the process of generating scientific knowledge and the ways in which it is initiated and maintained' (Rheinberger 2010, p. 2).

27 See for example Gingras (2010, p. 442) and Feest and Sturm (2011, p. 288).

28 My main reference in this respect is Chimisso (2003; 2008), who provides the fundamental background needed to account for the emergence of Bachelard's and Canguilhem's reflections.

29 Chimisso 2008, pp. 91–93. Bréhier considered bibliographic entries an example of data that the historians should collect.

30 Bréhier's idea occurs in the unsigned editorial of the first issue of *Revue d'histoire de la philosophie* and is therefore attributable to its director Emile Bréhier (see

Anonymous (1927), quoted in Chimisso 2003, p. 303). For Brunschvicg's take on the issue see L. Brunschvicg *et al.* 'Histoire et Philosophie. Séance du 31 mai 1923', p. 162, quoted in Chimisso 2008, p. 73.

31 For another take on the normative turn of French historical epistemology see the chapter by Simons contained in this volume.

32 Canguilhem 2005, p. 200. This passage can be read as a reformulation of Abel Rey: 'la théorie de la connaissance n'est qu'une idéologie vague ou une dialectique verbale, sans l'histoire philosophique de la science' (Rey 1935, p. XVIII). With these words, Rey meant to convey an aversion for any philosophy of science intended as a theory of science (*Erkenntnistheorie*) or a general methodology of science. To some extent this is a modulation or echo of the Kantian motto ('Thoughts without content are empty, intuitions without concepts, blind'), which Hanson (1962) and Lakatos (1971) later rephrase ('Philosophy of Science without History of Science is empty, History of Science without Philosophy of Science is blind').

33 Canguilhem 1974, p. 66.

34 Canguilhem 2005, pp. 202–203.

35 Ibid.

36 Canguilhem 2005, p. 203. As an instance of such a discourse Canguilhem quotes Hélène Metzger's *La genèse de la science des cristaux* (1918).

37 Ibid., p. 200.

38 Ibid. Particularly striking here is the contrast with Laudan (cf. footnote 18 *infra*).

39 Ibid.

40 Ibid.

41 Canguilhem 1968, p. 180. This translation is mine.

42 Canguilhem 2005, p. 200. Rheinberger, in his historical survey on the primarily twentieth-century process that brought epistemology to hybridise with history, claims that 'Historical epistemology has its own permanent laboratory in the past and future history of the sciences' – thus obscuring the normative metaphor of the tribunal. Rheinberger (2010, p. 66) misinterprets Canguilhem's use of the metaphor of the laboratory, assuming he is subscribing to the position of Eduard J. Dijksterhuis, the Dutch historian of science whose version of the metaphor Canguilhem discusses in his 1966 talk. As we have just shown, Canguilhem reports this image as one of the two possible ways of doing history of science, the other being the tribunal, or court, where the historian plays the role of the judge. It is this latter image that Canguilhem explicitly endorses. Rheinberger's reading has produced some distortions, including Schickore's (Schickore 2011; Arabatzis and Schickore 2012) depiction of Canguilhem as an anti-normative historicist-hermeneutist. This is clearly not the case, since, as we shall see, Canguilhem's historical epistemology is thoroughly normative.

43 See Chimisso 2001, Ch. 3 in which Bachelard's rationalism is read through the teacher-pupil dialectic.

44 Bachelard 1951, p. 25.

45 Canguilhem 2005, p. 201.

46 Chimisso rightly points out that Bachelard's and Canguilhem's respective approaches to the normative turn show remarkable divergences, in part as a consequence of the specific sciences to which they are applied (Chimisso 2003, 2015). For the sake of the argument developed here, I am more interested in the continuities between them than in their particularities.

47 Canguilhem 1955, pp. 60–69.

48 I borrow the idea of a 'triumphalist' history of science from Chang (2009), about which I will say more later.

49 Canguilhem 1955, pp. 166–167.

50 See Hull 1993, Pitt 2001, and Burian 2001.

51 Chang exemplifies his claim with reference to the historiography of chemistry and in particular to the standard accounts of the abandonment of the phlogiston theory and the success of Lavoisier's chemical revolution.
52 Chang 2009, p. 251.
53 Chang 2009, p. 253.
54 For Cohen's anti-presentist worries, see Cohen (1974).
55 See footnote 42 above.
56 Canguilhem 1988, p. 12.
57 Canguilhem 1981, p. 24: 'An ideology is ... a knowledge as far from its given object as it thinks itself bound to it. Here ideology would be the misunderstanding of the fact that any knowledge with a critical grasp of its project and its problem knows from the start that it is at some distance away from its operationally constituted object'.
58 On the necessity to reassess presentism in the light of French historical epistemology, see Braunstein 2008 and Loison 2016.

Bibliography

Arabatzis, Theodore, and Jutta Schickore, 'Introduction: Ways of Integrating History and Philosophy of Science', *Perspectives on Science*, 20. 4(2012), 395–408.

Bachelard, Gaston, *L'activité rationaliste de la physique contemporaine* (Paris: Puf, 1951).

Bird, Alexander, 'The Historical Turn in Philosophy of Science', in Stathis Psillos and Martin Curd (eds.) *The Routledge Companion to Philosophy of Science* (London & New York: Routledge, 2008), 67–77.

Braunstein, Jean-François, 'Les trois querelles en histoire des sciences', in *L'histoire des sciences. Méthodes, styles et controverses*, ed. by Jean-François Braunstein (Paris: J. Vrin, 2008), 87–103.

Burian, Richard M., 'The Dilemma of Case Studies Resolved: The Virtues of Using Case Studies in the History and Philosophy of Science', in *Perspectives on Science*, 9. 4(2001), 383–404.

Canguilhem, Georges, *La formation du concept de réflexe aux XVIIe et XVIII siècles* (Paris: J. Vrin, 1955).

Canguilhem, Georges, *Études d'histoire et philosophie des sciences* (Paris: J. Vrin. 1968).

Canguilhem, Georges, 'Gaston Bachelard', *Scienziati e tecnologi contemporanei* (Milano: Mondadori, 1974).

Canguilhem, Georges, *Ideology and Rationality in the History of the Life Sciences* (Cambridge, MA: MIT Press, 1988).

Canguilhem, Georges, 'What is Scientific Ideology?', *Radical Philosophy*, 29(1981), 20–25.

Canguilhem, Georges, 'What is the object of the history of sciences', in *Continental Philosophy of Science*, ed. by Gary Gutting (London and New York: Routledge, 2005), 198–207.

Chang, Hasok, 'We Have Never Been Whiggish (About Phlogiston)', *Centaurus*, 51 (2009), 239–264.

Chimisso, Cristina, *Gaston Bachelard: Critic of Science and the Imagination* (London: Routledge, 2001).

Chimisso, Cristina, 'The Tribunal of Philosophy and its Norms: History and Philosophy in Georges Canguilhem's Historical Epistemology', *Studies in History and Philosophy of Biomedical and Biological Sciences*, 32(2003), 297–327.

Chimisso, Cristina, *Writing the History of the Mind* (London: Ashgate, 2008).

Chimisso, Cristina, 'Narrative and Epistemology: Canguilhem's Concept of Scientific Ideology', *Studies in History and Philosophy of Science*, 54(2015), 64–73.

Cohen, Bernard I., 'History and the Philosopher of Science', in *The Structure of Scientific Theories*, ed. by Frederick Suppe (Urbana: University of Illinois Press, 1974), 308–361.

Damerow, Peter, 'The Origins of Writings as a Problem of Historical Epistemology', *Cuneiform Digital Library Journal*, 1(2006), 1–10.

Davidson, Arnold I., *The Emergence of Sexuality: Historical Epistemology and the Formation of Concepts* (Cambridge, MA: Harvard University Press, 2001).

Domski, Mary, and Dickinson, Michael, eds, *Discourse on a New Method: Reinvigorating the Marriage of History and Philosophy of Science* (Chicago: Open Court, 2010).

Feest, Uljana and Sturm, Thomas, eds, What (Good) is Historical Epistemology?, special issue of *Erkenntnis*, 75. 3(2011), November.

Galison, Peter, 'Kuhn and the Quantum Controversy', *British Journal for the Philosophy of Science*, 32(1981), 71–85.

Giere, Ronald N., 'History and Philosophy of Science: Intimate Relationship or Marriage of Convenience?', *British Journal for the Philosophy of Science*, 24. 3(1973), 282–297.

Giere, Ronald N., *Explaining Science: A Cognitive Approach* (Chicago: The University of Chicago Press, 1988).

Gingras, Yves, 'Naming Without Necessity: On the Genealogy and Uses of the Label "Historical Epistemology"', *Revue de synthèse*, 6th ser., 131. 3(2010), 439–454.

Hacking, Ian, 'Historical Meta-Epistemology' in *Wahrheit und Geschichte*, W. Carl and L. Daston (eds.) (Göttingen: Vandenhoek and Ruprecht, 1999), 53–77.

Hacking, Ian, 'Historical Ontology' in *Historical Ontology*, ed. by Ian Hacking (Cambridge, MA: Harvard University Press, 2002), 1–26.

Hanson, Norwood Russell, 'The Irrelevance of History of Science to Philosophy of Science' *The Journal of Philosophy*, 59(1962), 574–586.

Hull, David, 'Testing Philosophical Claims about Science' in *PSA: Proceedings of the Biennial Meeting of the Philosophy of Science Association*, II: *Symposia and Invited Papers*, ed. by D. Hull, M. Forbes, and K. Okruhlik (East Lansing: Philosophy of Science Association, 1993), 207–219.

Kitcher, Philip, 'Epistemology without History is Blind', in Feest, U. and Sturm, T. (eds.), *What (Good) is Historical Epistemology?* Special issue of Erkenntnis 75 (2011), 505–524.

Kuhn, Thomas S., *The Structure of Scientific Revolutions* (Chicago: University of Chicago Press, 1962).

Kuhn, Thomas S., *The Essential Tension: Selected Studies in Scientific Tradition and Change* (Chicago: The University of Chicago Press, 1977).

Krüger, Lorenz, 'History and Philosophy of Science: A Marriage for the Sake of Reason', in *Proceedings of the VI. International Congress for Logic, Methodology and Philosophy of Science*, ed. by L. J. Cohen, J. Łoś, H. Pfeiffer & K.-P. Podewski (Amsterdam: Elsevier, 1982), 108–112.

Krüger, Lorenz, 'Does a Science need Knowledge of its History?' [1978] in *Why Does History Matter to Philosophy and the Sciences? Selected Essays by Lorenz Kruger*, ed. by T. Sturm, W. Carl & L. Daston (Berlin: Walter de Gruyter, 2005), Ch. V.1.

Lakatos, Imre, 'History of Science and Its Rational Reconstructions.' In *Boston Studies in the Philosophy of Science*, vol. viii, ed. by Roger C. Buck, and Robert S. Cohen (Dordrecht: D. Reidel, 1971), 91–108.

Lalande, André, *Vocabulaire technique et critique de la Philosophie (1902–1923)* (Paris: Puf, 1972).

Laudan, Larry, 'Progress or Rationality? The Prospects for Normative Naturalism', *American Philosophical Quarterly*, 24. 1(1987), 19–31.

Laudan, Larry, 'Thoughts on HPS: 20 Years Later', *Studies in History and Philosophy of Science*, 20(1989), 9–13.

Laudan, Larry, and Rachel Laudan, 'The Re-Emergence of Hyphenated History-and-Philosophy-of-Science and the Testing of Theories of Scientific Change', *Studies in History and Philosophy of Science*, 59(2016), 74–77.

Loison, Laurent, 'Forms of Presentism in the History of Science: Rethinking the Project of Historical Epistemology', *Studies in History and Philosophy of Science*, 60 (2016), 29–37.

Mauskopf, Seymour, and Tad Schmaltz, eds, *Integrating History and Philosophy of Science: Problems and Prospects* (Berlin: Springer, 2012).

Pitt, Joseph C., 'The Dilemma of Case Studies: Toward a Heraclitean Philosophy of Science', *Perspectives on Science*, 9. 4(2001), 373–382.

Poovey, Mary, *A History of the Modern Fact: Problems of Knowledge in the Sciences of Wealth and Society* (Chicago: University of Chicago Press, 1998).

Rey, Abel, 'Avant-propos', *Thales*, 1(1935), xv–xix.

Rheinberger, Hans-Jörg, *On Historicizing Epistemology* (Stanford: Stanford University Press, 2010).

Schickore, Jutta, 'More Thoughts on HPS: Another 20 Years Later', *Perspectives on Science*, 19. 4(2011), 453–481.

6 Obligation to judge or judging obligations: The integration of philosophy and science in Francophone Philosophy of Science

Massimiliano Simons

Introduction: What French epistemology can offer iHPS

At the beginning of the 1970s, the Hungarian-born philosopher of science Imre Lakatos (1981, p. 107) famously claimed: 'Philosophy of science without history of science is empty; history of science without philosophy of science is blind.' In the wake of this slogan, integrated History and Philosophy of Science (iHPS) has often been understood in two ways. Firstly, Philosophy of Science (PS) needs History of Science (HS) for its content. A good illustration of this is the plea made in the 1980s by authors such as Ian Hacking or Peter Galison for the necessity of a 'philosophy of experiment'. These authors argued through specific historical case studies that PS must be enriched by a second dimension: science is more than just theory, it also includes a specific logic of experimentation, which 'has a life of its own' (Hacking 1983, p. 150).

Secondly, PS can also be deemed necessary for HS because certain discussions about science cannot be resolved through purely empirical means. Take the debate between realism and constructivism, a practical example of which is whether DNA has an independent existence or is rather a product of scientific research. Philosophers such as Alan Nelson (1994) have argued that such questions cannot be resolved by historical means only, but also need philosophical work.

Both these examples show that iHPS is warranted. Nevertheless, thinkers such as Larry Laudan (1989, p. 11) have concluded that 'historically based philosophy of science often remains more a slogan than a reality'. More recently Jutta Schickore (2011) has argued that the major reason for this unsuccessful integration is the 'confrontational model' that 'has dominated the debates about iHPS until today' (p. 462). Schickore here is referring to how in iHPS, historical case studies are merely seen as testing grounds for abstract philosophical theories, and she argues that such a model is 'highly problematic, best to be abandoned' (p. 456). Instead she argues for a hermeneutical approach, starting from the idea that PS always already implies a historical perspective. iHPS should be about this interconnectedness of PS and HS.

The aim of this chapter is to show how Francophone PS, or what is called French (historical) epistemology, embodies this interconnectedness. Moreover, a novel approach to what constitutes French epistemology will be developed here, going beyond a purely historical survey or a re-evaluation of a range of concepts found in this tradition.[1] The aim is instead to highlight two methodological principles at work in French epistemology that are often in tension with one another, but are not recognised as such in the literature. This will be done in the first section of the chapter by highlighting some general characteristics of French epistemology and subsequently elaborating these two principles. First of all, there is a *primacy of science over philosophy:* for French epistemologists scientific practices, and not philosophy, should provide the relevant categories by which these practices can be understood. Secondly, at the same time, French thinkers feel the obligation to make a normative judgement about the history of science.[2] The foundation for such normative judgements remains disputed, especially in the light of the first principle.

In the second section of this chapter, this tension will be illustrated by one of French epistemology's central figures, namely Gaston Bachelard. Both principles are present in his work: the primacy of science over philosophy in his 'surrationalism', with his normative judgements about the history of science being present in his distinction between lapsed and sanctioned history. The third and final part of this chapter will further argue that these principles are still at work in contemporary Francophone philosophers such as Michel Serres, Bruno Latour and Isabelle Stengers, a fact neglected within the literature. Their work must be understood as still being loyal to the principle of the primacy of science, but as dealing with the normative question in a rather different way. This novel approach, that I will call 'judging obligations', will be illustrated through the work of Stengers.

As stated before, the goal is not solely to give a historical overview, but to show how these ideas are useful for contemporary discussions. First of all, the principle of the primacy of science can be a very productive tool to develop a hermeneutical approach to iHPS. Secondly, the question of normativity will be readdressed. Authors within iHPS have often noted that there is a tension between a descriptive HS and a normative PS.[3] For this dilemma, Stengers offers a novel way out, in which one can both aim for comprehensive historical descriptions and nevertheless allow normative judgements about scientific practices.

6.1 French historical epistemology: an overview

6.1.1 The primacy of science over philosophy

French epistemology can be seen as a distinct way of integrating PS and HS, different from those in the Anglo-American or analytic tradition. Traditionally, one associates PS with attempts to formulate specific criteria for scientificity. Often the aim is to conceptualise a timeless model of science, namely a model that would work for any specific moment in time. Moreover, analytic PS tries to create a norm for scientific practice, i.e. to specify a way in which science *should* behave rather than dealing with how it actually behaves. In this sense PS has the task of dictating to the scientist how to do science. It is precisely such an *a priori* approach that many scholars in iHPS find problematic and to which they aim to offer an alternative.

In France, however, the philosophical landscape is completely different. There has never been a real distinction between PS and HS that has needed to be bridged in the first place. Rather, in French epistemology, PS and HS have always been integrated in a very particular way. There are several explanations for this. Firstly, there are institutional reasons. For instance, the first French chairs and institutions devoted to HS were founded by philosophers such as Gaston Milhaud (1858–1918) and Abel Rey (1873–1940). Moreover, French philosophy students have always had to combine their philosophical studies with a scientific discipline.[4]

Secondly, there is the intellectual legacy of Auguste Comte (1798–1857), who claimed that the only significant way to do epistemology is through HS. The latter is thus chosen as a method for non-historical goals: if French epistemologists write histories of science, it is never to focus on finding out specific historical details, but in order to 'write the history of the mind'.[5] The French epistemologist Léon Brunschvicg (1869–1944) illustrates this by refusing to label himself as a historian, despite having written several historical studies about physics and mathematics, and instead claiming that his task was 'not to know the *nature of things*, but to tell how *the human mind* works'.[6]

Brunschvicg contrasted his own 'philosophy of thought' with traditional PS. According to him, the latter was the attempt to flesh out a general framework for the conditions of the possibility of thought with the result of merely imposing on scientific thinking some historically situated categories. His own philosophy of thought, on the other hand, followed 'the course of thought with all the twists and sharp turns, the steady lakes and rapid waterfalls, the natural rivers'.[7] Brunschvicg's contemporary, Émile Meyerson (1859–1933), made a similar distinction between 'philosophy of nature' and 'philosophy of the intellect'. For Meyerson specific historical facts are not relevant as such, since he 'only deals with them as indicators of the role they played in reasoning'.[8]

At first sight it might seem strange that history could be of any help in understanding the human mind since the latter is traditionally the object of psychology. French epistemologists were, however, deeply critical of psychology's methods. They were especially skeptical towards introspective epistemology, where the mind looks inward and tries to understand itself by self-reflection. According to French epistemologists there is no reason to believe that the mind has immediate access to its own content.

In his *Course positiviste*, for instance, Comte (1998, p. 80) refuted the 'psychological method' by which the mind 'pretends to accomplish the discovery of the laws of the human mind by contemplating it in itself'. For him, this kind of enquiry only leads to biases and prejudices. The only 'scientific' way to understand how knowledge is produced is by looking at its history, namely HS. Similarly, Meyerson (1936, p. 206) claimed that reason 'does not know itself, because of the simple fact that it cannot observe itself. All one could know of its proper functioning, can only be concluded through the analysis of its products.'

From this perspective a different idea of philosophy was developed. Philosophy should not impose its categories on scientific practices but must respect their historical fluidity. One of the clearest examples of this is Jean-Toussaint Desanti's (1914–2002) *La philosophie silencieuse* (Desanti 1975) in which the author criticised several forms of PS. Predating Schickore's diagnosis by almost forty years, Desanti dismissed a PS that would reduce scientific theories to mere illustrations of philosophical questions. According to Desanti, such a PS translates scientific practices into philosophical questions that, allegedly, only philosophers would be able to ask and answer. Desanti sketches a whole history of such philosophical reductions of science, or what he calls an 'interiorisation' of science by philosophy, be it an interiorisation in the *Eidos* (Plato), the human understanding (Descartes, Spinoza), the subject (Kant), the concept (Hegel) or consciousness (Husserl). They all ignore what Desanti called *une philosophie silencieuse* (a silent philosophy), the autonomous philosophical productivity of scientific practices themselves: 'Within the philosophical field organized in such a way, the sciences end up (in the eyes of the philosophers, to be clear) having no existence but the one attributed to them by the questions posed by the philosophers'.[9] According to the traditional view, scientists know nothing about their own practices and need PS to articulate their truth. For Desanti, however, such a perspective ignores that scientists themselves have the capability of producing philosophically interesting categories and distinctions.

One of the central principles of French epistemology is thus, what I will call the *primacy of science over philosophy:* the idea that philosophy should not impose its categories on the history of science but should follow its open and productive movement. In this sense, French epistemologists mobilise HS to problematise the idea of a fixed set of philosophical categories to grasp the nature of science.

6.1.2 The question of normativity

This idea of the primacy of science over philosophy, however, often comes into conflict with another constant in French epistemology, namely the question of normativity. Comte and Meyerson both recognised, for example, the shifting history of the mind, while stressing that, although the products of science are diverse, there is nonetheless an all-encompassing principle at work. For Comte this principle was the law of the three stages, for Meyerson the principle of identity: in both cases, the plurality of the world must be united in one law.[10] Thus, although they criticised traditional PS for imposing its own norms on the history of science, at the same time both introduced a normative criterion by which they judged history.

Other French epistemologists remained more loyal to the primacy of science, by stressing the openness of the history of science. Brunschvicg, for instance, dismissed introspective epistemology not merely because of epistemological doubts, but also for ontological reasons. Reason is, according to him, ontologically defined as a dynamic process, and therefore always shifting throughout its history. It is for this reason that Brunschvicg, although deeply inspired by Kant, strongly criticised his predecessor's fixed framework of categories. For Brunschvicg (1922, p. 550), 'we are more faithful than Kant to the spirit of critical idealism if we reject the table of univocal categories in order to follow the dynamism and plasticity of intellectual functions'.

To understand such an ontological claim, another important characteristic of French epistemology has to be taken into account besides the institutional setting and Comte's legacy mentioned in the introduction: namely, the intellectual context that existed at the beginning of the twentieth century. French epistemologists working at that time were confronted with foundational crises in mathematics and physics.[11] It is not necessary to discuss the origins or development of these crises here, since what is important is solely how philosophers understood them to be undermining the traditional foundations for rationality. How can we still be sure that our beliefs are rational if there can be historical breaks in our understanding, such as the discovery of non-Euclidean geometry or the theory of relativity?

One could interpret projects like Logicism or Husserlian phenomenology as attempts to recreate a new firm foundation for rationality. Some French epistemologists, however, took a different approach. Rather than trying to find a firm foundation for all scientific revolutions, authors such as Brunschvicg claimed that rationality resided within the revolutionary act itself. Historicity is not seen as a problem for rationality, but precisely as its ground: science is rational not despite, but because of its historical shifts. The same tension led André Lalande (1867–1963), another French epistemologist, to the distinction between 'a *constituted reason* and a *constitutive reason*'.[12] Behind the superficial discontinuities of constituted fields of reason lies a constant normative constitutive reason. Moreover, the discontinuities are precisely the proof of the continuous rational action of this constitutive reason.

Brunschvicg and Lalande thus mobilised the primacy of science in a novel way, namely by invoking it against the normative principles of other French epistemologists, such as Comte or Meyerson. The history of science is not guided by an atemporal goal, but rather, is subjected to an 'indefinite progress'.[13] This tension between the primacy of science and normativity is at the core of many debates within French epistemology. This is especially clear in the work of Brunschvicg's student, Gaston Bachelard (1884–1962). In the next section Bachelard's 'surrationalism' will be explored in more detail. Firstly, because Bachelard is one of the most influential authors within French epistemology; secondly, because it is his work that authors like Serres and Latour later criticised.

6.2 The philosophy of Gaston Bachelard

6.2.1 Bachelard's surrationalism

In Bachelard's oeuvre, the principle of the primacy of science over philosophy is present under the banner of his 'open rationalism', or what he terms *surrationalism*.[14] For Bachelard, rationalism cannot boil down to a series of all-encompassing fixed categories which determine how we think and act. He contrasted this 'closed rationalism' with his own 'open rationalism', which stressed that the act of rationality consisted in overcoming predetermined categories of thought by creating new ones. Bachelard wanted 'to place reason *inside the crisis*, to prove that the function of reason is *to provoke crises*'.[15] Similarly to the subversive nature of surrealism, *surrationalism* aimed to break with the conservatism of closed rationalism and become an *avant-garde* rationalism by grasping the novel ways in which rational thinking occurs within contemporary sciences. For Bachelard 'science instructs reason. Reason has to obey to science, a more evolved science, an evolving science.'[16]

This openness Bachelard was looking for is, however, not located in traditional philosophy but instead within the sciences themselves. At the centre of his work is therefore a tension between philosophy and science. For him, the sciences continually revise their theories, while philosophy tends to be conservative about them, seeing them as atemporal and universal. Like Desanti, Bachelard claimed that philosophers often try 'to apply a necessarily finalist and closed philosophy to open scientific thought'.[17] Such PS does not recognise that 'science ordains philosophy by itself'[18] and that 'contemporary science is philosophical at its core'.[19] The primacy of science implied articulating the internal philosophical categories at work within scientific practices, rather than imposing those made up by philosophers.

This idea was also at work in his famous notion of the *epistemological rupture* 'between ordinary knowledge and scientific knowledge'.[20] This rupture implies a break with ordinary imagination, but also with spontaneous philosophical theories about science. 'The scientific mind precisely consists in the bracketing of the philosophy one starts with. Just as experimental activity, philosophy linked to scientific activity must be nuanced and, as a consequence, be mobile.'[21] But traditional PS does not do this, so according to Bachelard's famous words, 'science does not have the philosophy it deserves'.[22] And this is precisely what Bachelard aimed to create in his own work, an open rationalism that respects the primacy of science over philosophy:

> It would therefore be interesting, we believe, to understand scientific philosophy on its own, to judge it without preconceived ideas, outside, moreover, the too strict obligations of traditional philosophical vocabulary. In fact, science creates philosophy. Philosophers must therefore adjust their own language in order to translate the contemporary [scientific] thinking in all its flexibility and mobility.[23]

To really grasp what is going on in the scientific practices, Bachelard argued it is necessary to look at the history and development of these sciences. Thus, Bachelard exemplifies the extent to which, for French epistemology, an iHPS is the only way to understand scientific practices. Against problematic abstract philosophical reflection, Bachelard (1953, p. 223) proposed 'that it is at the level of particular examples that the philosophy of science can give us general lessons'.

6.2.2 Bachelard and the obligation to judge

Using the example of Bachelard, I have shown what the principle of the primacy of science entails. This element of Bachelard's work has not received the attention it deserves within secondary literature and has therefore been stressed here. What has, however, often been discussed is how the second principle is present in his work, namely the question of normativity.

As many scholars have noted, a central element of Bachelard's philosophy is his claim that it is always necessary to make normative judgements within HS, while taking present-day science as the normative criterion.[24] This stance is often called 'presentism', though Bachelard's version of it is of a specific kind.[25] He contrasts the work of the epistemologist to that of the historian. The historian searches for and accumulates facts without making normative judgements. This model, however, does not work for HS, because 'it does not take into account that every historian of science is necessarily a historiographer of Truth'.[26] The element of truth brings in a specific normative element that other domains of history lack. This is a strong claim and not widely shared in contemporary HS. A further analysis of why French epistemologists such as Bachelard nevertheless made such a claim is therefore necessary, and can best be elaborated through one of Bachelard's pupils, Georges Canguilhem (1904–1995).[27]

In his lecture 'The object of the history of science', Canguilhem articulated the normativity at stake by contrasting the object of science and the object of HS. The object of science can be considered as a 'natural object', for instance, crystallography studies crystals, and biology studies organisms. The object of HS, however, is a series of statements about these natural objects, statements made within a particular culture. In this sense, the objects of science and of HS are different, and are respectively that of a natural object, and that of a cultural object. According to Canguilhem this leads to at least five further differences between their objects. First of all, the

> object of historical discourse is, in effect, the historicity of scientific discourse, inasmuch as this historicity represents the carrying out of an internal law-governed project, but one which is traversed by accidents, retarded or deflected by obstacles, interrupted by crises, i.e. moments of judgements and of truth.

Secondly, the object of HS is something that evolves, and concerns 'an object to which incompleteness is essential'.[28] HS cannot be the natural history of a given object with a fixed identity, but raises the question of the identity of the object within history.

Thirdly, it follows that 'the object of the historian of science can only be delimited by a decision which assigns it its interest and importance'.[29] The historian is therefore in need of a norm to determine what to include and exclude as part of the science he or she wants to examine. Fourthly, as a result of this delimitation, it also has a necessary connection to realms which are typically considered non-scientific: ideology and society that themselves have to be defined by the enquiry. The object of HS is therefore not self-contained, but must be constructed. Finally, to transgress a mere 'chronological register', HS must be seen as an 'axiological activity, the search for truth'.[30] To become more than a mere list of scientific instruments, texts and statements, in order to grasp the normative force of scientific arguments, HS has to become normative itself. One has to grasp the specific field of concepts, theories and ideas of a certain period, which implies a certain normative choice. For Canguilhem, such a choice has to be made in light of present-day findings within the scientific discipline of which the examined part of history is seen as *its* history.

For similar reasons Bachelard considered the history of physics and chemistry as implying a normative perspective, which judged historical episodes in the light of the present. This did not mean that such episodes were seen as necessary steps with the present as their goal, but rather that the present always rewrote its own history from this necessity of normativity. For Bachelard this resulted in a distinction between 'lapsed history' (*histoire périmée*) and 'sanctioned history' (*histoire sanctionnée*).[31] He saw the former as the parts of science that, from a contemporary perspective, were excluded as non-science, while the latter referred to the preserved elements.

Such a harsh distinction must moreover be situated within Bachelard's broader philosophy. First of all, Bachelard endorsed the goal of French epistemology to write a history of the mind. The aim was therefore to grasp the normative force of reason through its history, not the historical details. Secondly, for Bachelard there was a clear pedagogical task present in his philosophy.[32] Describing the struggle of past science with certain epistemological obstacles, such as ordinary experience or naïve philosophical theories, allowed present students of science to be wary of possible missteps and confusions. HS, for Bachelard, was thus about the formation of the scientific mind. So-called Whiggish history of science fails in this objective, since it tends to either present contemporary theories as self-evident – and thus forgets the danger of certain epistemological obstacles that prevent many from fully understanding these theories – or reproduces certain naïve images – such as seeing atoms as solar systems – and thus slowing down the dynamics of the scientific mind, rather than freeing it from these images.

6.3 From epistemological to political normativity

6.3.1 A new generation of Francophone philosophers of science

In the previous sections, I showed how French epistemologists conceived the relationship between PS and HS. I showed what the principle of the primacy of science consisted in: a principle that can be useful for developing an approach to iHPS that avoids the confrontational model, but promotes what Schickore calls a 'hermeneutical' approach. Their views on normativity, however, might not be shared by contemporary iHPS, but nevertheless offer a strong case of why a certain normativity within iHPS can still be warranted.

More recent developments in Francophone PS have often been neglected in overviews of French epistemology, but in fact offer an even more promising normative project. Michel Serres, Bruno Latour and Isabelle Stengers, though in several ways opposed to French epistemology, share some of its fundamental principles, while drawing radically different normative conclusions from them.

To relate them to French epistemology is not self-evident, since their projects have several fundamental differences, and they are often explicitly in opposition to French epistemology.[33] Serres, for instance, states:

> Yes, I wrote my thesis under Bachelard, but I thought privately that the 'new scientific spirit' coming into fashion at that time lagged way behind the sciences. ... The model it offered of the sciences could not, for me, pass as contemporary. This new spirit seemed to me quite old. And so, this *milieu* was not mine.[34]

But, in this passage Serres is criticising Bachelard for specific reasons. Serres is not claiming that there have been no relevant shifts within the history of science, nor that such shifts would be unworthy of philosophical attention. Rather, he claims that Bachelard is lagging behind the newest developments, since in mathematics for instance, 'instead of speaking of algebra, topology, and set theory, [Bachelard] referred to non-Euclidean geometries, not all that new'.[35] Serres accused Bachelard of not being loyal enough to the principle of the primacy of science by missing out on the most recent scientific developments. According to Serres, therefore, Bachelardian epistemology was still not open enough.[36]

For Serres, these new scientific developments also resulted in a greater reflexivity within the sciences about their own philosophical categories. The consequence of such an increased reflexivity was a new tension within the self-identity of the epistemologist. If the sciences became capable of articulating their own philosophical categories, then the traditional French epistemological project was reduced to a mere repetition of what scientists did themselves. In what sense, then, did the 'philosopher's work differ from that of a journalistic chronicler, who announces and comments on the news?'[37] Either the epistemologists merely repeated the sciences, or their projects aimed to do something more, but this extra element would involve a form of philosophical speculation, going beyond what the sciences said. Such speculation thus implied a tension within the principle of the primacy of science.

Moreover, Serres claimed that Bachelard had always done more than mere description. There had always been a speculative-normative project at work, namely the above described presentism, looking for epistemological obstacles and epistemological ruptures. For someone like Serres, the project of French epistemology was a crypto-normative project, starting from a philosophical model of purification: true science purified itself from all the obstacles, from imagination, from myth. Against this Serres stated that 'there is no purer myth than the idea of a science purified from all myth'.[38]

This criticism was even clearer in the work of Serres' pupil, Bruno Latour. In *Laboratory Life*, Latour criticised Bachelard's epistemological project, especially the claim that science was rational in so far it succeeded in breaking with ordinary experience and imagination.[39] Latour observed no such break in his fieldwork in scientific laboratories. For him, notions such as irrationality were never genuine descriptions of science, but rather accusations that aimed to discredit an opponent.[40] A genuine understanding of science should not start from such unquestioned distinctions as rational versus irrational.

Latour's criticism was partly inspired by the Strong Programme of the Sociology of Scientific Knowledge (SSK).[41] But it is equally important to note that Latour went further than these sociologists. This led to a different appreciation of what to do with such distinctions as rational/irrational or true/false. Rather than ignoring or bracketing them, as SSK does,

> our contention is not just that sociological explanation should be impartial with respect to truth and falsity, and that both sides of the dichotomy require explanation. Our argument is that the implicit (or explicit) adaption of a truth value alters the form of explanatory account which is produced.[42]

For Latour, although such distinctions could not be used as explanations, they were nevertheless real by virtue of their effects in the scientific field. Since such distinctions were found within scientific practices, often at moments of controversy, they should be taken into account as well. Latour thus invoked the principle of the primacy of science by arguing that, since these distinctions were used within scientific practices, we should therefore track how scientists used them and not just dismiss them as ideology.

Moreover, Latour's work can be seen as a further development of this principle, by stressing the necessity to follow *all* forms of associations and connections one finds in the scientific practices. Latour, for instance, stated that

> Instead of defining *a priori* the distance between the nucleus of scientific content and its context, an assumption that would render incomprehensible the numerous short-circuits between ministers and neutrons, science

studies follows leads, nodes, and pathways no matter how crooked and unpredictable they may look to traditional philosophers of science.[43]

The scientific actors should decide which distinctions are relevant and which are not, even if it leads us 'outside' of science, and philosophers

> should be as undecided as the various actors we follow as to what tech-noscience is made of; to do so, every time an inside/outside division is built, we should follow the two sides simultaneously, making up a list, no matter how long and heterogeneous, of all those who do the work.[44]

This is Latour's famous dictum to 'follow the actors', which can be interpreted in the light of the primacy of science over philosophy. A similar rule can be found in the work of Isabelle Stengers, who wants to give a description of the scientific practices that 'does not insult the scientists'.[45]

6.3.2 Judging the obligations of the scientists

Serres, Latour and Stengers were quite close to the earlier generation as far as the principle of the primacy of science is concerned, but they did not follow Bachelard's normative project. In fact, similarly to authors such as Brunsch-vicg before them, they played out the principle of the primacy of science against the normative project of their predecessor. Precisely Bachelard's ideas about how scientific rationality always excludes imagination, and how such a distinction should be mirrored in HS as lapsed versus sanctioned history, prevented him from truly endorsing the primacy of science. According to this new generation, in order to follow the sciences, the whole way through, the philosopher should not be the judge over what is and is not part of science.

Does this mean that philosophers should give up any ambition to make normative judgements about scientific practices? Although it might seem so, I would argue that this is not the main message. The conclusion I wish to draw from these recent criticisms is that the problem is not the fact that Bachelard judges, but the manner in which he judges. What was at stake for this new generation was not the normativity of Bachelard's project per se, but its implicit endorsement of a certain political project: one that made a distinction between science, which knows and as such is epistemologically superior to non-science, which does not.[46] This implied a problematic reaffirmation of the political power of the sciences over the rest of society. One should make an analytical distinction between the epistemological project of French epistemology that endorses the primacy of science and their political project of granting science hierarchical autonomy from other practices. What authors such as Serres, Latour or Stengers had in mind was a different political project: one that respected the primacy of science – follow the actors, do not insult the scientists – but at the same time conceptualised a new politics of science.[47]

It is this project that can provide another useful tool for contemporary iHPS, as illustrated by the work of Stengers. Similar to Latour and Serres, she agrees that there is no essential distinction between science and non-science; she nevertheless argues that such a distinction is *constructed* by the actors in the field. This distinction relies, for her, on two elements, namely a range of *requirements* and a range of *obligations*.[48] For Stengers the essential element to be respected is the obligations, while the requirements can be the object of a normative judgement. A different politics of science is therefore possible, namely by replacing the requirements, while respecting their obligations, and thus the primacy of science.

In short, a *requirement* refers to the fact that some distinctions are deemed necessary in order for a practice to work. Think for instance of the distinction between science and non-science, or science seen as autonomous from external socio-economic history. In the current scientific practices these distinctions are claimed as requirements, for, otherwise, scientists would allegedly lose themselves in discussions with their financial sources, who would impose their own goals, or with the general public, who might want to have a say in the development of science and its innovations. Many contemporary scientists argue that, in order to function, scientific practices require these distinctions to be maintained, even if they are constructed.

However, these requirements are not enough on their own. Scientific practices are also shaped by their specific *obligations*. This is related to the idea that scientists do not get involved in scientific controversies because they choose to. Rather they feel *obligated* to by the phenomena or the problem itself. What scientists find interesting and worth pursuing is not something they freely decide themselves, but rather is partly caused by the phenomena that spark their interest. Scientists thus feel a certain obligation being imposed from the outside, from the phenomena they try to take into account. Scientists are then not speaking in their own name, but in the name of these phenomena. An *obligation* thus refers to the fact that there is a relation that makes a difference: things with or without taking the phenomena into account are not the same.

For Stengers, such obligations are at work in all types of practices and can be of many sorts. For instance, within religious practices an obligation can consist in Virgin Mary who obliges believers to take Her into account by, for example, asking them to go on a pilgrimage.[49] Stengers stresses that her 'project does not thereby seek to ground a privilege for the sciences, which alone would escape sociological analysis. The same type of questions should be posed with regard to other practices.'[50] In the case of the experimental practices, scientists are obligated to speak with their phenomena in mind. In the same way a pilgrimage is experienced as being imposed from outside the believer, scientists feel obligated to make sure everything necessary is done in order for their phenomena to be properly accounted for in scientific discussions. It is this obligation towards the phenomena that is at the core of the scientific practice.

Scientists, therefore, have certain obligations and act as they do in light of them. But although these obligations are experienced by scientists, their nature remains opaque. The success of answering to these obligations, allowing a certain scientific practice to function, depends on what Stengers calls an 'event'. An event refers to the situation in which a scientific practice has found a way to properly articulate its phenomena. But again, the nature of this success remains opaque and does not grant the participants to this event a privileged explanation of why this approach is successful. The crucial point is that one can make a separation here between two elements, namely one can aim 'no longer to deny the differences scientists claim for themselves, but to avoid any way of describing them which implies that scientists have a privileged knowledge of what this difference that singularizes them *signifies*'.[51]

According to Stengers, both philosophers and sociologists of science falsely assume that if such an event has occurred, it must be transparent to its participants. Not only philosophers who try to define general criteria of rationality, but also sociologists who deny that such criteria exist, make the mistake of assuming that if such criteria cannot be properly articulated by the scientists doing the research that no such event has occurred. 'According to the viewpoint I am defending, the scope of [such] demonstration is zero, for it assumes that the foundational event can give an account of itself.'[52] Scientists know they have obligations, but this does not mean that what these are is fully transparent to the scientists. As Stengers writes, '[j]ust as the event, in itself, does not have the power to dictate how it will be narrated or the consequences that will be authorized on its behalf, neither does it have the power to select among its narrators'.[53] The consequence of this is that there is no guarantee that the specific 'requirements' scientists demand to safeguard the success of the event are necessary. Requirements can thus be disputed without disputing that obligations exist.

This opens up the possibility to come up with other requirements, for instance a range of requirements free from a strong distinction between science and non-science or open to contributions from citizens to scientific research without immediately compromising its scientificity. What Stengers therefore proposes is a project in which specific scientific disciplines can be politically judged, without dismissing the primacy of science. 'This is why my position clearly does not amount to *defending the sciences*, but to defending their singularity in order to utilize it to invent the means of a critical position that *complicates* their history.'[54] According to Stengers, social constructionist HS, such as SSK, went too far by confusing requirements with obligations and dismissing them both. Scientists are correct in protesting against such historians who disrespect their obligations, but that does not entail that their proposed requirements are indisputable. Rather, these are always open for debate.

Conclusion: new tools for iHPS

This chapter shows that within the tradition of French epistemology and its critics, a very specific integration between science and philosophy is conceptualised, a relationship wherein any form of primacy of philosophical speculation over scientific practices is fiercely disputed. Rather, the idea is that scientific practices produce their own philosophical categories and thus a primacy of science over philosophy. The task of French epistemologists has been, therefore, to start from these elements, instead of dictating what is philosophically relevant to science. At the same time, it was made clear that a constant struggle in French epistemology remains present, where this primacy of science conflicts with the normative ambitions of French epistemologists.

This conflict was mainly shown through the work of Bachelard, creating a tension between his surrationalism and his obligation to judge the history of science. This is precisely the object of the critique of contemporary authors such as Serres, Latour and Stengers. However, their criticisms are not a dismissal of French epistemology, but can be understood as a particular radicalisation of this primacy of science over philosophy. Bachelard and others were not wrong in their objectives, but failed to follow them the whole way through – there were still traces of philosophical *a prioris* in their work, resulting in the endorsements of certain political projects related to the social status of science. In that sense, the first conclusion is that to speak of a radical break between these two traditions is false. Rather one must speak of a radicalisation or a revision of the 'primacy of science'.

This chapter, however, aimed for more than a historical overview of recent developments within Francophone philosophy of science. It precisely aimed to highlight possible productive tools for contemporary discussions of how PS and HS should be integrated. First, the principle of the primacy of science embodies the 'hermeneutical' perspective called for by contemporary authors such as Schickore. This essential principle of French epistemology can thus offer guidance of how such a hermeneutically-inspired iHPS should be pursued.

Secondly, this chapter addressed the question of normativity, central to many discussions in iHPS. How can a descriptive HS and a normative PS be combined? Rather than giving up any normative ambitions, French epistemology indicates several avenues through which normative elements still play a role. Not only is the 'obligation to judge' the past from the present, embodied in the work of Bachelard and Canguilhem, worth discussing, but the particular project of 'judging obligations', found in the work of Stengers, offers a similarly promising avenue for a normative iHPS.

From this perspective, there is room in iHPS for normative judgements about scientific practices. Not only in the activity of conceptually separating the obligations from their requirements, but even in a critical evaluation of the history of these obligations. Either iHPS can have the task of showing how certain obligations are under threat by developments within society, such as the rise of the contemporary knowledge economy and its focus on applications and economic gain. Or, iHPS can focus on contemporary shifts within the sciences, where certain obligations are being forgotten, badly articulated, or where different requirements for them can be conceived. But a constant in Francophone philosophy of science is the claim that scientists are at least seen as potential allies in this reform and that philosophy must be open enough to learn from them.

Notes

1 See respectively Cristina Chimisso (2008) and Hans-Jörg Rheinberger (2010).
2 This is especially clear once contrasted with Thomas Kuhn. Kuhn was heavily criticised by French epistemologists for what they saw as his lack of attention to the normative element at work within the history of science. See Massimiliano Simons (2017a), pp. 41–50.
3 Ronald Giere (1973), p. 290.
4 Cristina Chimisso (2008); Anastasio Brenner (2015).
5 Cristina Chimisso (2008).
6 Léon Brunschvicg (1922), p. xiii (own translation).
7 Ibid., p. 570.
8 Émile Meyerson (1936), p. 118 (own translation).
9 Jean-Toussaint Desanti (1975), p. 68 (own translation).
10 Comte claimed that all sciences followed a historical three-stage pattern, starting from a theological stage (explanation by divine causes), subsequently going through a metaphysical stage (explanation by abstract philosophical categories), to end up in the positive stage (explanation by laws instead of causes). Meyerson disagreed and argued that the search for causality is central to the human mind, in the sense that scientists always look for an identity relation between antecedents and consequents, expressed in a formal equation. Meyerson explained shifts in the history of science as different attempts and approaches to find such an identity relation.
11 See Enrico Castelli Gattinara (1998).
12 André Lalande (1963), p. 17 (own translation).
13 Léon Brunschvicg (1922), p. 595.
14 See Gaston Bachelard (1940).
15 Gaston Bachelard (1972), p. 27 (own translation).
16 Gaston Bachelard (1940), p. 144 (own translation).
17 Ibid., p. 2.
18 Ibid., p. 22.
19 Gaston Bachelard (1953), p. 180 (own translation).
20 Ibid., p. 207.
21 Gaston Bachelard (1951), p. 17 (own translation).
22 Gaston Bachelard (1953), p. 20.
23 Gaston Bachelard (1934), p. 3 (own translation).
24 For instance Cristina Chimisso (2001).
25 Laurent Loison (2016).
26 Bachelard (1953), p. 86.
27 For methodological differences between the two, see Cristina Chimisso (2015).
28 Georges Canguilhem, (2005), p. 203.
29 Ibid., p. 203.
30 Ibid., p. 204.
31 Gaston Bachelard (1951), 'Chapitre 1'.
32 Cristina Chimisso (2001).
33 First of all they are critical of the notion of 'rationality'. Secondly, they start from a fundamental different ontology. See Massimiliano Simons (2017b).
34 Michel Serres and Bruno Latour (1995), p. 11.
35 Ibid.
36 Especially the early work of Serres aims to update Bachelard's work. See Massimiliano Simons, (2018, forthcoming).
37 Serres and Latour (1995), p. 15.
38 Michel Serres (1974), p. 259 (own translation).
39 Bruno Latour and Steve Woolgar (1986), pp. 151–153.
40 Bruno Latour (1987), p. 185.

41 This programme, associated with the so-called Edinburgh School, aimed to apply a sociological analysis to scientific knowledge. See David Bloor (1976).
42 Latour and Woolgar (1986), p. 149n1.
43 Bruno Latour (1999), p. 99.
44 Latour (1987), p. 176.
45 Isabelle Stengers (2006).
46 Isabelle Stengers (2000), p. 26.
47 Massimiliano Simons (2017b).
48 Isabelle Stengers (2010), p. 52.
49 Isabelle Stengers (2006).
50 Isabelle Stengers (2000), p. 58.
51 Ibid., p. 67.
52 Ibid., p. 68.
53 Ibid., p. 68.
54 Isabelle Stengers (1997), p. 143.

Acknowledgements

The research for this chapter was supported by the Flandres Research Foundation (FWO). Further I would like to thank Matteo Vagelli and Eugenio Petrovich for making the chapter possible, and Katleen Pasgang, Charlotte Alderwick, and the editors of this volume for their many useful comments and suggestions.

Bibliography

Bachelard, Gaston, *Le nouvel esprit scientifique* (Paris: Alcan, 1934).
Bachelard, Gaston, *La philosophie du non* (Paris: PUF, 1940).
Bachelard, Gaston, *L'activité rationaliste de la physique contemporaine* (Paris: PUF, 1951).
Bachelard, Gaston, *Le matérialisme rationnel* (Paris: PUF, 1953).
Bachelard, Gaston, *L'engagement rationaliste* (Paris: PUF, 1972).
Bloor, David, *Knowledge and Social Imagery* (London: Routledge and Kegan Paul, 1976).
Brenner, Anastasio, 'Is There a Cultural Barrier Between Historical Epistemology and Analytic Philosophy of Science?', *International Studies in the Philosophy of Science*, 29. 2(2015), 201–214.
Brunschvicg, Léon, *L'expérience humaine et la causalité physique* (Paris: Alcan, 1922).
Canguilhem, Georges, 'The object of the history of science', in *Continental Philosophy of Science*, ed. by Garry Gutting (Oxford: Blackwell, 2005), 198–207.
Castelli Gattinara, Enrico, *Les inquiétudes de la raison* (Paris: Vrin, 1998).
Chimisso, Cristina, *Gaston Bachelard: Critic of Science and the Imagination* (London: Routledge, 2001).
Chimisso, Cristina, *Writing the History of the Mind* (Aldershot: Ashgate, 2008).
Chimisso, Cristina, 'Narrative and epistemology: Georges Canguilhem's concept of scientific ideology', *Studies in History and Philosophy of Science*, 54(2015), 64–73.
Comte, Auguste, *Auguste Comte and Positivism: the Essential Writings*, ed. by Getrud Lenzer (New Brunswick: Transaction, 1998).
Desanti, Jean-Toussaint, *La philosophie silencieuse ou Critique des philosophies de la science* (Paris: Seuil, 1975).

Giere, Ronald, 'History and Philosophy of Science: Intimate Relationship or Marriage of Convenience? (Book Review)', *The British Journal for the Philosophy of Science*, 24. 3(1973), 282–297.

Hacking, Ian, *Representing and Intervening* (Cambridge: Cambridge University press, 1983).

Lakatos, Imre, 'History of Science and Its Rational Reconstructions', in *Scientific Revolutions*, ed. by Ian Hacking (New York, NY: Oxford University Press, 1981), 107–127.

Lalande, André, *La raison et les normes* (Paris: Hachette, 1963).

Latour, Bruno, *Science in Action* (Cambridge, MA: Harvard University Press, 1987).

Latour, Bruno, *Pandora's Hope: Essays on the Reality of Science Studies* (Cambridge, MA: Harvard University Press, 1999).

Latour, Bruno and Steve Woolgar, *Laboratory Life: the Construction of Scientific Facts* (Princeton: Princeton University Press, 1986).

Laudan, Larry, 'Thoughts on HPS: 20 years later', *Studies in History and Philosophy of Science*, 20. 1(1989), 9–13.

Loison, Laurent, 'Forms of Presentism in the History of Science. Rethinking the Project of Historical Epistemology', *Studies in History and Philosophy of Science*, 60 (2016), 29–37.

Meyerson, Émile, *Essais* (Paris: Vrin, 1936).

Nelson, Alan, 'How Could Scientific Facts be Socially Constructed?', *Studies in History and Philosophy of Science*, 25. 4(1994), 535–547.

Rheinberger, Hans-Jörg, *On Historicizing Epistemology: An Essay* (Stanford: Stanford University Press, 2010).

Schickore, Jutta, 'More Thoughts on HPS: Another 20 Years Later', *Perspectives on Science*, 19. 4(2011), 453–481.

Serres, Michel, *Hermès III. La traduction* (Paris: Éditions de Minuit, 1974).

Serres, Michel and Bruno Latour, *Conversations on Science, Culture and Time* (Ann Arbor: Michigan University Press, 1995).

Simons, Massimiliano, 'The Many Encounters of Thomas Kuhn and French Epistemology', *Studies in History and Philosophy of Science*, 61(2017a), 41–50.

Simons, Massimiliano, 'The Parliament of Things and the Anthropocene: How to Listen to "Quasi-Objects",' *Techné: Research in Philosophy and Technology*, 21. 2–3 (2017b), 150–174.

Simons, Massimiliano, 'Surrationalism after Bachelard: Michel Serres and le nouveau nouvel esprit scientifique', *Parrhesia*, 31(2019, forthcoming).

Stengers, Isabelle, *Power and Invention: Situating Science* (Minneapolis: University of Minnesota Press, 1997).

Stengers, Isabelle, *The Invention of Modern Science* (Minneapolis: University of Minnesota Press, 2000).

Stengers, Isabelle, *La vierge et le neutrino* (Paris: Seuil, 2006).

Stengers, Isabelle, *Cosmopolitics I* (Minneapolis: University of Minnesota Press, 2010).

Part 2
iHPS in practice

7 Experimentalist as spectator: The phenomenology of early modern experimentalism

Mark Thomas Young

Introduction

This chapter focuses on the theme of practical knowledge in order to contribute to the growing body of literature that explores the relationship between craft traditions and the emergence of experimental science. As will be shown, attending to the role ascribed to practical contexts and goals in the production of experimental knowledge reveals important differences between the practices of experimentalists and artisans that encourage us to reconsider the extent to which craft traditions can be understood to have influenced the development of experimental methods in early modern science. Besides adding an important dimension to the question concerning the relationship between craft and experimental natural philosophy, the focus on practical knowledge will also be shown to yield insights into another, equally difficult object for historical analysis: the nature of experience in experimental demonstrations. In order to explore this relationship between practical knowledge and experience, this chapter will integrate elements in the History of Science with a Philosophy of Science that is informed by the existential phenomenology of Martin Heidegger. As we will see, Heidegger's early work highlighted the role that practical contexts play in determining how phenomena are experienced, and for this reason, represents a valuable supplement to existing accounts of experience in early modern science.[1] In this way, the integrated approach this chapter adopts will highlight aspects of experimental practice that are often overlooked by conventional historical scholarship. It is a central claim of this chapter then that it is the failure to explore the phenomenological context of experimental practices that has helped lead scholars towards overstating the continuities between craft practices and experimentalism.

My argument will proceed in three parts: the first section will briefly outline the historiography of the conventional division between craft based and theoretical knowledge. The second section explores the central role accorded to practice in the epistemology of early modern craftwork. The third section of the chapter will utilise a phenomenological perspective to show how experimentalists' attempts to distance themselves from practical contexts led to the development of epistemological practices that were based around forms of experience fundamentally different from those central to craft traditions. Whereas practice and practical goals formed a constitutive aspect of the epistemology of craft, early experimentalists were instead committed to the idea that natural knowledge is best acquired outside of practical contexts, a presupposition which will be shown to have structured both their own investigations into nature and their claims to be able to improve existing artisanal practices.

7.1 Historiographical issues surrounding the divide between craft based and theoretical knowledge

Since antiquity, it has been common to distinguish the knowledge employed by the artisan (*techne*) from the theoretical knowledge of the scholar (*scientia*). In classical Greece, this distinction took two main forms: the first is reflected in Aristotle's hierarchy of knowledge, which underscored the epistemological benefits of theory over productive forms of knowledge;[2] the second distinction occurred within the arts themselves and concerned the ends towards which they were employed. For Aristotle, an art was 'banausic' or vulgar:

> If it renders the body or soul or mind of free men useless for the employments and actions of virtue. Hence we entitle vulgar all such arts as deteriorate the condition of the body, and also the industries that earn wages; for they make the mind preoccupied and degraded.[3]

These distinctions persisted throughout the early middle ages, where operative and theoretical forms of knowledge and practice were often kept separate from one another.[4] For while it was not uncommon to find positive descriptions of craftwork in medieval classifications of knowledge, universities still excluded technical practice from their curriculums while intellectuals continued to regard the practical and manual aspects of artisanal practice with suspicion.

Beginning in the late middle ages however, university educated scholars began to show an increased interest in craftwork. The spread of humanist culture during the Renaissance, for example, often encouraged the celebration of practical knowledge and the skilful achievements of the craftsman.[5] Emboldened by the increased scholarly interest in their work, craftsmen began authoring technical treatises that represented their work in both written and pictorial forms. Furthermore, changes to the economic and political climate in Europe throughout this period created new contexts in which technical skills came to be seen as valuable by an increasingly wide range of actors. The skills required to build fortifications attracted the attention of monarchs, for example, while mechanical toys found receptive audiences in the courts of princes.

This apparent rehabilitation of the social status of the craftsman has long been viewed as a significant factor in explaining the emergence of experimental science in the early modern period. During the seventeenth century a new group of natural philosophers began to subvert established traditions of natural philosophy by arguing for the epistemological superiority of craft practices. Their claims that scientific and technological progress could only be achieved by drawing upon the knowledge and methods of craft practitioners have often been taken to signify the culmination of the rising status of technical practitioners in early modern Europe while at the same time marking the origins of experimental natural philosophy. Together, these observations form the basis of the received view, which understands modern science to have emerged from a combination of aspects of craft practice and natural philosophy made possible by the rising status of the artisan.[6] In support of this view, historians have often drawn attention to various ways in which artisanal traditions influenced experimentalism, such as methodological values including 'the use of instruments; the practices of direct observation and experimentation; methods of precise measurement and other forms of quantification; and a positive valuation of individual experience'.[7]

Yet while this has long represented a central element of the historiography of the scientific revolution, the extent and manner in which artisanal culture influenced the emergence of experimental science is itself the subject of ongoing dispute.[8] For we should not overlook the fact that early experimentalists also went to great lengths to establish the superiority of their activities over those of craftsmen. As we will see, seventeenth century experimental treatises often reveal a complex balance between positive valuations of craft and attempts to denigrate both the methods by which artisans worked and the knowledge which they had hitherto achieved.

7.2 Craftwork and the primacy of practice

The concept of craft has a tumultuous history, ebbing and flowing to encompass different activities in response to changes in intellectual culture and the structure of society. Pre-modern Europe, for example, operated with a wider conception of craftwork than is common today. In the late medieval period, the meaning of the term 'artisan', often extended beyond those who skilfully manufactured goods for sale and included those who provided both unprocessed products, such as fishmongers, as well as those providing services, such as innkeepers.[9] At the same time, craftwork was commonly understood under the rubric of the 'mechanical arts' (*artes mechanicae*), a term which emerged in the early Middle Ages to denote a variety of activities including among others, hunting, cooking, building and agriculture.[10] While early modern intellectuals often continued to employ this wide definition of craft and artisanship, two changes are worth noting. Firstly, the increased popularity of mechanical devices and theory throughout the Renaissance led to a strengthening of the association between the mechanical arts and the original Greek meaning of *mechanica* as the mathematical analysis of moving objects. Secondly, an increasing scholarly interest in esoteric traditions throughout the Renaissance, alongside their eagerness to be associated with craft,[11] helped expand the meaning of *artes mechanicae* to include natural magic and alchemical practices.

Lately, this diverse set of practices has received increased attention from historians who, following the recent turn towards material culture in the history of science, examine the history of craft from a variety of different disciplinary orientations. The growing body of literature that has resulted from such studies has not only given rise to an increasingly detailed picture of craft, but also highlighted the possibility of making general claims about early modern craft knowledge by revealing the extent to which traditions of practice as diverse as alchemy and sculpture often shared the same epistemological presuppositions.[12]

One of the more universal features of the epistemology of craft to be revealed by such studies is the emphasis on the primacy of practice in acquiring knowledge concerning natural materials or technical processes. The tendency to draw on a dichotomy between practice and theory, and then to emphasise the epistemological benefits of the former, is common in Renaissance technical treatises. For instance, the Italian goldsmith Cellini, who in detailing instructions for soldering, is careful to remind the reader that: 'it is practice and experience, together with a man's own discretion, that are the only real ways of teaching one how to bring about good results in this or in anything'.[13]

In technical treatises covering a wide variety of crafts, early modern authors often acknowledged the limitation of written words to convey the knowledge necessary for the practice of a craft. This was due not only to the inherent difficulties of codifying the judgment of practitioners,[14] but also to the nature of the work itself. In detailing procedures used for melting metals, for example, the sixteenth century Italian metallurgist Vannoccio Biringuccio notes that:

> it is necessary for you to understand well what you wish to do in this operation [for] having chosen with good judgment the way in which you are to proceed, you will easily arrive where you intend. But because the light of judgment cannot come without practice, which is the preceptress of the arts, I shall pass through this briefly with the idea of one day being able to supplement this further by demonstrating it to you.[15]

Here we see how the benefits of demonstration over general rules in the attempt to acquire craft knowledge were often understood to stem from the inherent variability in the qualities of the objects artisans worked with. This was a view found not only in the work of literate craftsmen, but also in the work of natural philosophers such as the English experimentalist Robert Boyle who, in detailing a technique for writing without ink, is careful to add that: 'a few trials will teach better than the rules (because according to the goodness and calcination of the Vitriol, the proportion of the other ingredients must be varied)'.[16]

Variation affected both the materials with which the artisan worked and the wider environment within which they operated. Fluctuations in temperature and humidity, for example, required adjustments in a wide variety of activities, including bread-making, painting and casting, while changes in military technology required engineers to continually modify the designs of fortresses. A central element of craft skill consisted therefore in the development of intuitive judgment through lengthy periods of practice by which the practitioner could deal with the variation of materials in a particular environment. For this reason, technical treatises often provided more information concerning how one should go about developing individual judgment than useful instructions for actually performing certain tasks. Take Biringuccio's account of assays, procedures aimed at determining the contents or quality of an ore or metal, as an example; beyond providing general descriptions of the equipment used for smelting and schematic images of laboratories, little information is given concerning how to go about actually performing an assay.[17] What we find instead are suggestions for different methods that may help a practitioner judge the quality of an ore themselves, such as methods of experimentation, coupled with a repeated emphasis of the central role of judgment in assaying practices and the dangers associated with inexperience. The inadequacy of the text as an instructional manual for practice, is not unusual. As Stephan Epstein has observed, this was a common characteristic of pre-modern technical literature, the authors of which:

> seldom practised what they described and so typically overestimated the role played by explicit propositional knowledge in craft and engineering practice. Written manuals were incomplete and sometimes misleading; they might contain technical details not actually applied in solving the problem; and they left out crucial practising 'tricks'.[18]

The limitations of early modern technical literature as guides for practice have led some scholars, such as Pamela H. Smith to suggest that the actual function of literary genres such as books of secrets and craft recipes were as 'mechanism[s] for asserting the validity of natural knowledge gained by practice'.[19] This importance to which authors of technical treatises ascribed practice in the early modern period highlights a number of distinctive features of the epistemology of craft. Below I will discuss three of the most important features for the purposes of this essay: first, the embodied acquisition of knowledge; second, the way properties of materials were conceptualised; and finally, the communication of craft knowledge.

Firstly, the relation between the acquisition of natural knowledge and the body of the practitioner. In many crafts, the development of manual dexterity alongside an ability to use one's fingers, ears, eyes or nose to discern subtle differences in materials was the result of the experience gained in performing operations upon natural or artificial materials. In the epistemology of craft then, the body is central not only in the act of acquiring knowledge, but also stands as a repository for the knowledge gained; for experience comes to be stored in the body as it is configured through the repetitive performance of certain tasks.

Secondly, the way the properties of materials were often indexed to the goals of practitioners. Technical treatises reflect how a large part of learning a craft involved developing the ability to recognise properties of natural materials that are most useful for particular tasks. In the *Pirotechnica*, Biringuccio spends ample time detailing the properties of charcoal; woods with an earthy property, for example, yield strong charcoal which is suitable for 'operations that need long, live and powerful fires'.[20] Furthermore, it was not only naturally occurring properties that were conceived in relation to the productive activity of the artisan, but also properties that practitioners would cultivate themselves. Cellini, for example, details a method for increasing the 'fattiness' of clay to be used in casting, and notes that 'this particular kind of fattiness in no wise hinders the accepting of the metal, indeed it accepts it infinitely better, and the clay holds a hundred times more firmly'.[21]

Finally, the primacy of practice highlights the particular modes by which craft knowledge was transmitted. As is well known, craft practices were learned primarily through face to face interactions in a variety of settings, including the family household,[22] and worksites where practitioners could be exposed to alternative methods of practice. The importance of these modes of transmission is reflected in the central role played by apprenticeships in the training of artisans in a wide variety of fields. While considerable variation existed in the education provided by premodern apprenticeships among, and even within occupations,[23] certain structural features appear to have remained standard. Among the most important for our purposes here is that the process of learning a craft was nearly always integrated with productive activity.[24] All stages of an apprenticeship involved contributing in some way to the services provided by the master; apprentices would typically begin with menial tasks such as cleaning before gradually progressing to more complicated and demanding forms of productive activity. As formal instruction was minimal, mastering each stage depended on the apprentice themselves acquiring experience through trial and error. As Bert De Munck and Hugo Soly note:

> In nearly all branches of industry, apprentices needed to learn how the raw materials would react to the mechanical and chemical production processes, in a context of variable surrounding conditions, such as temperature, level of humidity, quality of the materials used, and other elements that were often impossible to measure accurately. The only way to understand these processes was to do the work and to keep evaluating the reactions. Didactic skills were of secondary importance: masters were merely expected to point out what had gone wrong and what might be improved, while apprentices, in turn, had only to acquire sufficient practical experience.[25]

Developing this kind of expertise required great lengths of time; in England, for example, a minimum apprenticeship period of seven years was dictated by the Statute of Artificers (1563).

As we will see, the lengths of time deemed necessary for the development of these skills were viewed as one of the major shortcomings of artisanal education by early modern natural philosophers.

7.3 Heidegger and the epistemology of early experimentalism

Our exploration of the central role of practice in apprenticeships provides us here with a natural opportunity to turn to the work of philosopher Martin Heidegger, who posited a fundamental relation between practical contexts and experience. One of the central goals of Heidegger's early work was to challenge the epistemological priority that western culture has long attributed to disinterested contemplation. To do so, Heidegger distinguishes two different ways in which we can experience the world around us.

The first is a form of understanding that Heidegger called *zuhandenheit* (ready-to-hand), which results from manipulating things in the midst of practical projects. Here Heidegger notes how when we perform various tasks, the things we use appear for us both in light of the goals that they can help us achieve, and the wider context of objects and practices which are required for their successful employment.[26] It is because things appear to us in this way when they are being used, as equipment (*das Zeug*), that they have a tendency to recede from our attention. In using a hammer, to draw on Heidegger's own example, our attention is focused not upon the entity in our hand but rather on the task of erecting the frame of the house by driving nails into timber. Rather than appearing as a self-sufficient entity in its own right then, the hammer is experienced as part of a wider backdrop of equipment that also includes nails and timber. Furthermore, the same applies for the subject doing the experiencing; when using equipment, we are often absorbed into the task we perform and therefore do not experience ourselves as distinct subjects interacting with an external world. Instead, this way of interacting with the world is experienced as a form of 'absorbed coping' in which our subjectivity becomes interwoven with the activity performed.

The second mode of experience, *vorhandenheit* (present-at-hand), arises when things in the world are experienced *outside* of practical contexts. In this mode of experience, entities do not withdraw into the background but instead appear for us as discrete objects that possess a certain set of properties. According to Heidegger, this mode of experience characterises the stance adopted by the scientist who deliberately adopts a theoretical attitude which attempts to comprehend entities in isolation from the ordinary contexts of practice in which they usually appear. This scientific stance reveals entities as *vorhanden*, that is to say, as objects with definite properties that are available for analysis by a distinct subject. For example, the hammer appears to us in this mode of understanding as possessing:

> the 'property' of heaviness: it exerts a pressure on what lies beneath it, and it falls if this is removed. When this kind of talk is so understood, it is no longer spoken within the horizon of waiting and retaining an equipmental totality and its involvement-relationships. What is said has been drawn from looking at what is suitable for an entity, with 'mass'. We have now sighted something that is suitable for the hammer, not as a tool,

but as a corporeal Thing subject to the law of gravity. To talk circumspectively of 'too heavy' or 'too light' no longer has any 'meaning'; that is to say, the entity in itself, as we now encounter it, gives us nothing with relation to which it could be 'found' too heavy or too light.[27]

Heidegger was careful to note that scientific knowledge also emerges from the active manipulation of things in the world, suggesting in particular that not only practices of experimentation, but also observation, emerge through *praxis*.[28] Adopting the theoretical stance for Heidegger, therefore involves subtracting away particular ways in which entities are revealed to active and interested agents, and can therefore be understood to represent a derivative mode of engaging with the world:

> we *overlook* not only the tool-character of the entity we encounter, but also something that belongs to any ready-to-hand equipment: its place. Its place becomes a matter of indifference. This does not mean that what is present-at-hand loses its location altogether. But its place becomes a spatio-temporal position, a 'world-point', which is in no way distinguished from any other.[29]

The remainder of this section will explore the extent to which Heidegger's conception of a theoretical attitude corresponds to the stance early modern experimentalists maintained towards the natural world. Doing so will highlight important differences between the forms of experience privileged by craft traditions and experimentalism that have often been overlooked by traditional scholarship.

At a glance, the received view that the emergence of experimentalism involved the legitimation of craft practices appears to find widespread support in the literature of early modern experimental philosophy, where it is common to find claims of the epistemological superiority of the 'man of experience' over the 'man of learning'. However, it is also often overlooked how the project of undermining the authority of scholastic natural philosophy by highlighting the superiority of craft practice, itself created a host of new problems for experimental philosophers. For, as Pamela H. Smith notes, 'because natural philosophers resembled artisans in their interests, they were often anxious about making clear the differences in their social and intellectual status'.[30] Fellows of the Royal Society often went to great lengths to preserve such differences; despite the Society's emphasis on the benefits of practical experience for example, only a few tradesmen managed to overcome the opposition from existing fellows to become members in the latter stages of the seventeenth century.[31] Furthermore, while many fellows of the society followed Bacon in praising the achievements of craft over that of the philosophy of the schools, few if any considered the state of contemporary trades to be adequate. To the contrary, it was common to characterise the mechanical arts as a field of technical knowledge that remained far from realising its full potential. Royal Society fellow William Petty, for example, described the current state of technological knowledge as:

like a field where a battle hath beene lately fought, where we may see many leggs, and armes, and eyes lying here and there, which for want of a union and a soule to quicken and enliven them, are good for nothing but to feed Ravens; and infect the aire. So we see many Wittes and Ingenuities lying scattered up and downe the world, whereof some are now labouring to doe what is already done, and pushing themselves to reinvent what is already invented, others we see quite stuck fast in difficulties, for want of a few Directions.[32]

What was lacking, according to Petty, was the guidance of experimental philosophy: a new approach to the study of nature upon which a comprehensive reform of natural and technological knowledge was to be based. In order to establish the superiority of experimental philosophy over existing practices in the mechanical arts, experimental philosophers drew upon traditional epistemological notions which had long been used to reinforce social barriers between manual workers and intellectuals. Chief among these is the classical notion of a banausic art, according to which a connection to practical contexts secures the social, spiritual, or epistemological inferiority of a form of knowledge. This idea informed early experimental philosophy in a variety of ways. We find it in the work of Francis Bacon, for example, who argues that technical progress can only be achieved by isolating the investigation of nature from practical contexts:

> the very abundance of mechanical experiments reveals the dearth of those that contribute and help most to informing the intellect. For the mechanic, caring nothing for investigating the truth, does not give his mind or reach out his hand to anything apart from what helps him in his work. But hope of further advancement of the sciences will be well grounded only when we take and gather into natural history many experiments of no use in themselves but which only contribute to the discovery of causes, which experiments I have grown used to calling *Light-bearing* as against *Fruit-bearing* ones. Now, the former have a marvelous virtue and quality to them, namely that they never fail or let you down. For since we bring them in not to accomplish any work but to show the natural cause in something, they suit their purpose whichever way they turn out because they settle the question.[33]

Bacon's criticisms of the epistemology of the craftsman proved to be remarkably influential. Successive generations of experimental philosophers routinely reiterated Bacon's claim that the connection to practice and practical goals confines the mechanic's attention only to that which appears immediately profitable while at the same time obscuring the significance and relevance of discoveries that are not considered to be related to their current project. Bacon's followers also extended this idea by outlining further reasons why the inherently practical nature of craftwork functioned to hinder technological progress. In *The History of the Royal Society*, for example, Thomas Sprat (1667, p. 391) notes how the lengthy and practical nature of the training a tradesman receives functions to restrict them to particular methods of working and prevent them from exploring new practices which might yield new technical discoveries. A more serious difficulty however, according to Sprat, lies in the tendency for craftsmen to 'jump the gun' by seeking to capitalise on discoveries before enough investigation has been performed to ascertain the true causes of a particular phenomenon:

> It diminishes that very profit for which men strive. It busies them about possessing some pretty Prize; while Nature itself, with all its mighty Treasures; slips from them: and so they are serv'd like some foolish Guards who, while they were earnest in picking up some small Money, that the Prisoner drop'd out of his Pocket, let the Prisoner himself escape, from whom they might have got a great randsom.[34]

Not surprisingly, such critiques of the epistemology of the artisan invariably served to outline a positive role for the experimental philosopher. Sprat continues the passage above, for example, by noting how this difficulty can be avoided by entrusting the investigation of nature to 'the care of such men, who by the freedom of their education, the plenty of their estates, and the usual generosity of Noble bloud, may be well suppo'd to be most averse from such sordid considerations'.[35] Far from attempting to conceal the homogeneity of the membership of the Royal Society then, Sprat purposely draws attention to the fact that 'the farr greater number are gentlemen, free and unconfin'd',[36] in order to underscore the epistemological benefits of investigating nature *outside* of practical contexts. According to Sprat, these benefits include possessing a more open mind for new discoveries,[37] and a psychological temperament that is more resistant to failure.[38] Even a lack of manual skill is suggested by Sprat to confer an advantage to the gentleman natural philosopher:

> they come to try those operations, in which they are not very exact, and so will be more frequently subject to commit errors in their proceeding; which very faults, and wanderings will often guide them into new light, and new conceptions.[39]

Sprat was not alone in regarding the manual dexterity of the craftsman as an obstruction in the path of technological progress. In *Some Considerations Touching the Usefulness of Experimental Philosophy*, Robert Boyle (1671) advanced a similar critique of the role of skill in the epistemology of craft. In particular, Boyle argued that the craftsman's reliance upon skill in their work at the expense of 'any diligent or accurate search into the qualities of those productions' has left many of the properties of natural materials undiscovered.

Boyle's work gives voice to a common contention among fellows of the early Royal Society, that contemporary artisanal practices were often radically inefficient:

> what the artificer undertakes, is either long in doing (as in the ordinary way of tanning, brickmaking, seasoning wood etc.) or takes up more paines, or requires a greater apparatus of instruments, or else is some other way more chargeable, or troublesome, or laborious to be effected than it needs to be.[40]

For some cases, Boyle maintained that 'things that are wont to be done by the labour of the hand, may with far more ease and expedition (the quantity considered) be performed by engines'.[41] Indeed, the development of mechanical devices that could replace the hand of the craftsman was a common goal of early experimentalists such as Hooke, who laboured to develop devices for astronomy and manufacture that would decrease the reliance of the natural philosopher on the artisan.[42]

Yet perhaps the central source of the inefficiency of craftwork, according to Boyle, was the cultivation of manual skills through lengthy apprenticeships, a feature of artisanal culture that he believed could often be rendered superfluous by the development of an experimental philosophy that provided 'knowledge of peculiar qualities, or uses of physical things (which) may enable a man to perform those things physically, that seem to require tools and dexterity of hand, proper to artificers'.[43] Like many of the fellows of the early Royal Society, Boyle envisaged such knowledge to take the form of a 'history of trades': a compendium of technical information gleaned from the observation of craft practices, which according to William Petty, would contain descriptions of:

> the whole Processe of Manual Operations and Applications of one Naturall thing (which we call the Elements of Artificials) to another, with the necessarie Instruments and Machines, whereby every peice of worke is elaborated, and made to be what it is, unto which work bare words being not sufficient, all Instruments and tooles must be pictured, and colours added when the discriptions cannot be made intelligible without them.[44]

The project of developing a history of trades represented the cornerstone upon which the experimental philosopher's claims to be able to improve the practice of craft depended. Petty, for example, expected a history of trades to reduce the time required for an apprenticeship in the trades by more than half.[45] Likewise, Thomas Sprat emphasised the capacity of such a work to help instruct:

> the worst artificers ... by considering the methods and the tools of the best: and [cause] the greatest inventors ... (to) be exceedingly inlighten'd; because they will have in their view the labours of many men, many places and many times, wherewith to compare their own.[46]

According to Sprat, invention was a project that was understood to be ideally suited to the gentleman, whose lack of rigorous professional training in any one discipline was suggested to have cultivated a 'large and unbounded mind' that was more capable of recognising undiscovered technical possibilities than the 'low and vulgar genius' of the craftsman.[47] Unfettered by practical interests, and armed with a history of trades, the experimental philosopher was thereby understood to attain a vantage point from which a wide variety of different trades could be assessed and compared simultaneously. This was understood to facilitate the recognition of, among other things, the potential for cross fertilisation – techniques unique to a particular art that might also prove beneficial to other forms of practice.

However, the great promise that experimental philosophy was believed to hold for the progress of technology depended not only on the separation of the knowing subject from practical contexts, but also upon the decontextualisation of the objects of knowledge. For before technical information could play such a role, it too had to be rendered independent of context. In the second volume of *The Usefulness of Experimental Philosophy*, Boyle (1671) envisions the task of the experimental philosopher as translating the expertise of the artisan on natural materials into context independent properties which could form the basis of an experimental science of materials:

> For ... although they commonly mean by such termes (of Goodness and Badness) no more, than the fitness, or unfitness of such things to yield a good price, and in order thereunto for the purposes they are to be imployed about in their particular Trades–, yet this fitness or unfitness is wont to consist in, or to suppose, Qualities, that may relate to divers other things, and be apply'd to many other purposes.[48]

In the same way that the testimony of the artisan was understood to require decontextualisation in order to realise its full epistemological potential, so too was the knowledge arising from experiment. For the majority of those who performed the experiments upon which the new science was to be based, such as Boyle's technicians in his laboratory, or even curators of the Royal Society such as Hooke, were paid for their services, invoking negative associations to the artisan, whose epistemological abilities were widely believed to be limited by association with personal interests and practical contexts. The solution was to separate the process by which experimental knowledge was generated from the performance of experiments themselves. For members of the early Royal Society, the production of experimental knowledge was conceived not as an operative task achieved by those performing the experiments, but as a discursive activity undertaken by the largely genteel membership, who would in a disinterested way:

> judge and resolve upon the matter of fact (by taking) an exact view of the repetition of the whole course of the experiment ... This critical and reiterated scrutiny of those things, which are the plain objects of their eyes; must

needs put out of all reasonable dispute, the reality of those operations, which the society shall positively determine to have succeeded.[49]

The manner in which experiments were to be selected for demonstration was also influenced by the ideal of disinterestedness. In an effort to avoid being misled by theoretical preconceptions, members of the society often aimed to separate the experience of experimental demonstrations from theoretical contexts altogether. According to Sprat, this meant that experiments were to be selected by a process in which members would:

> urge what came into their thoughts, or memories ... either from the observations of others, or from books, or from their own experience, or even from common fame itself. And in performing this, they did not exercise any great rigour of choosing, and distinguishing between truths and falsehoods: but a mass altogether as they came[50]

The result was an experimental programme which was notoriously eclectic. Among the various subjects addressed by the experimental agenda of the Royal Society were the investigation of properties of materials traditionally considered the purview of the artisan: the tensile strength of wire, for example, or the combustibility of different kinds of wood. However, demonstrations such as this were often performed on the same afternoon as anatomical dissections, or the presentation of chemical substances or mechanical devices that had been prepared earlier. This deliberately unstructured approach to experimental demonstration, coupled with the society's vigorous insistence on the practical benefits such a programme would yield, inevitably gave rise to parody. The most well-known example of a satirical depiction of the new scientific practices is Thomas Shadwell's *The Virtuoso* (1676). Here we find the difficulties facing the public legitimation of the Royal Society's approach to the investigation of nature reflected clearly in the exploits of the main character, Sir Nicolas Gimcrack, who in attempting to develop a method for swimming on dry land, proudly proclaims;

> I content myself with the speculative part of swimming;
> I care not for the practice. I seldom bring anything to use:
> 'Tis not my way. Knowledge is my ultimate end.[51]

Conclusion

The benefits of separating the acquisition of knowledge from practical contexts represent one of the central presuppositions underlying the experimental practices that emerged in England during the latter stages of the seventeenth century. Widely conceived as a prerequisite for technological progress itself, the decontextualisation of both the knowing subject and the object of knowledge constituted a guiding ideal which influenced the development of experimental practice and theory. Here we see how despite the repeated insistences of Society members to be breaking with established traditions and pursuing radically new avenues for research, this aspect of their thought instead reflected traditional epistemological assumptions which have held a central place in western philosophy since classical antiquity. Taking practical knowledge as our theme in this article has allowed us to see how, while far removed in terms of time and place, Heidegger and fellows of the early Royal Society nonetheless occupied a shared philosophical universe. For the same epistemological presuppositions that members of the society worked to affirm in practice and theory were precisely those that Heidegger was concerned to criticise through an exploration of the phenomenological relationship between practice and experience. In demonstrating how our experience of a dichotomy between subject and object emerges through the adoption of a stance of disinterested contemplation, Heidegger's thought suggests a fundamental difference between forms of experience common in craftwork and experimental practice. For unlike artisanal work which was structured by practical contexts, proponents of the new experimental practices approached nature as spectators, removed from their objects of knowledge through a set of practices designed to reap the epistemological benefits of disinterested contemplation.

Notes

1 See for example, Dear (2006) and the more recent Vanzo (2016).
2 Aristotle (1984), 981a30.
3 Aristotle (1932), 1337b.
4 Smith (2006), 17.
5 See for example, Rossi (1970) and more recently, Long (2011).
6 See for example, Rossi, (1970) and Hooykaas (1972).
7 Long (2011), p. 3
8 For recent accounts addressing this issue, see Cormack (2017) and my own Young (2018).
9 Cooper (2011), 6.
10 Whitney (1990), 60.
11 Epstein (2013).
12 See Smith (2006).
13 Cellini (1967), p. 12.
14 Epstein (2013), p. 52.
15 Biringuccio (2005), p. 280.
16 Boyle (1671) Essay 1, 32.
17 Biringuccio (2005), p. 135.
18 Epstein (2013), p. 52.
19 Eamon (2011), 33.
20 Biringuccio (2005), p. 174.
21 Cellini (1967), p. 113.
22 Long (2001), p. 73.
23 Lane (1996), p. 66.
24 Wallis (2008) and De Munck (2010).
25 De Munck and Soly (2007).
26 Heidegger (1962), p. 96 (68).

27 Ibid., p. 412 (361).
28 Ibid., p. 409 (358).
29 Ibid., p. 413 (362).
30 Smith (2006), p. 265.
31 Pumfrey (1991).
32 Petty (1648).
33 Bacon (2004) §99, 157.
34 Sprat (1667), p. 68.
35 Sprat (1667), p. 68.
36 Sprat (1667), p. 67.
37 'those men who are not particularly conversant about any one sort of arts, may often find their rarities and curiosities sooner than those who have their minds confin'd wholly to them', Sprat (1667), p. 391.
38 Sprat (1667), p. 392.
39 Sprat (1667), p. 392.
40 Boyle, 'That the Goods of Mankind May be much increased by the Naturalists insight into Trades' in Boyle (1671), p. 20.
41 Boyle (1671), p. 20.
42 See Illife (1995) and Burnett (2005) esp. chap. 3.
43 Boyle, 'Of Doing by Physical Knowledge what is wont to require Manual Skill' in Boyle (1671), 1 (205 ext).
44 Petty (1648), p. 18.
45 Petty (1648), p. 19.
46 Sprat (1667), p. 310.
47 Sprat (1667), p. 392.
48 Boyle, (1671), p. 6.
49 Sprat (1667), p. 99.
50 Ibid., p. 95.
51 Shadwell (1966), p. 47.

Bibliography

Aristotle, *Politics* (Loeb Classical Library) trans. H. Rackham (Cambridge: Harvard University Press, 1932).

Aristotle, *Metaphysics*, in *The Complete Works of Aristotle: The Revised Oxford Translation*, Vol. 1&2. W.D. Ross (trans.), Jonathan Barnes (ed.) (Princeton, NJ: Princeton University Press, 1984).

Bacon, Francis, *Novum Organum*, Graham Rees and Maria Wakely eds. *The Oxford Francis Bacon*, XI (Oxford: Clarendon Press, 2004).

Biringuccio, Vannoccio, (1540) *The Pirotechnica of Vannoccio Biringuccio: The Classic Sixteenth Century Treatise on Metals and Metallurgy*, eds. and trans. Cyril Stanley Smith and Marta Teach Gnudi (New York: Dover Publications, 2005).

Boyle, Robert, *Some Considerations Touching the Usefulnesse of Experimental Naturall Philosophy: the Second Tome containing the Later Section of the Second Part* (Oxford: Henry Hall, 1671).

Burnett, Graham D., *Descartes and the Hyperbolic Quest: Lens Making Machines and their Significance in the Seventeenth Century* (Philadelphia: American Philosophical Society, 2005).

Cellini, Benvenuto, *The Treatises of Benvenuto Cellini on Goldsmithing and Sculpture*, trans. C.R. Ashbee (New York: Dover, 1967).

Cooper, Lisa, *Artisans and Narrative Craft in Late Medieval England* (New York: Cambridge University Press, 2011).

Cormack, Lesley B., 'Handwork and Brainwork: Beyond the Zilsel Thesis', in *Mathematical Practitioners and the Transformation of Natural Knowledge in Early Modern Europe*, ed. by L. Cormack, S. Walton and J. Schuster (Cham: Springer, 2017).

Dear, Peter, 'The Meanings of Experience', in *The Cambridge History of Science Vol. 3: Early Modern Science*, ed. by Katharine Park and Lorraine Daston (Cambridge: Cambridge University Press, 2006).

De Munck, Bert, 'Corpses, Live Models and Nature: Assessing Skills and Knowledge before the Industrial Revolution (case: Antwerp)', *Technology and Culture*, 51. 2 (April 2010), 332–356.

De Munck, Bert and Hugo Soly, '"Learning on the Shop Floor" in Historical Perspective' in *Learning on the Shop Floor: Historical Perspectives on Apprenticeship*, ed. by Bert De Munck, Steven L. Kaplan, and Hugo Soly (New York: Bergham Books, 2007).

Eamon, William, 'How to Read a Book of Secrets' in *Secrets and Knowledge in Medicine and Science 1500–1800*, ed. by Elaine Long and Alisha Rankin (Farnham: Ashgate Publishing, 2011).

Epstein, S. R., 'Transferring Technical Knowledge and Innovating in Europe, c.1200 – c. 1800' in *Technology Skills and the Pre-Modern Economy in the West (Essays Dedicated to the Memory of S.R Epstein)* ed. by Maarten Prak and Jan Luiten van Zanden (Leiden: Koninklijke Brill NV, 2013).

Heidegger, Martin, *Being and Time*, trans. John Mcquarrie and Edward Robinson (Oxford: Basil Blackwell, 1962).

Hooykaas, Reijer, *Religion and the Rise of Modern Science* (Edinburgh: Scottish Academic Press, 1972).

Illife, Rob, 'Material Doubts: Hooke, Artisan Culture and the Exchange of Information in 1670's London', *The British Journal for the History of Science*, 28. 3(Sep., 1995), 285–318.

Lane, Joan, *Apprenticeship in England, 1600–1914* (London: UCL Press, 1996).

Long, Pamela, *Openness, Secrecy, Authorship: Technical Arts and the Culture of Knowledge from Antiquity to the Renaissance* (Baltimore: Johns Hopkins University Press, 2001).

Long, Pamela, *Artisan/Practitioners and the Rise of the New Sciences, 1400–1600* (Corvallis OR: Oregon State University Press, 2011).

Petty, William, *The Advice of W.P to Mr. Samuel Hartlib* (London, 1648).

Pumfrey, Stephen, 'Ideas Above his Station: A Social Study of Hooke's Curatorship of Experiments', *History of Science*, 29. 83(Mar 1, 1991), 1–44.

Rossi, Paolo, *Philosophy, Technology and the Arts in the Early Modern Era*, trans. by Salvator Attansio, ed. by Benjamin Nelson (New York: Harper & Row, 1970).

Shadwell, Thomas, *The Virtuoso*, ed. by Marjorie H. Nicolson and David S. Rodes (Lincoln: University of Nebraska Press, 1966).

Smith, Pamela, *The Body of the Artisan: Art and Experience in the Scientific Revolution* (Chicago: Chicago University Press, 2006).

Sprat, Thomas, *History of the Royal Society* (London: J. Martyn and J. Allestry, 1667).

Vanzo, Alberto, ed., *Experience in Natural Philosophy and Medicine* (Special Issue) *Perspectives on Science*, 24. 3(May-June 2016).

Wallis, Patrick, 'Apprenticeship and Training Premodern England', *The Journal of Economic History*, 68. 3(September 2008), 832–861.

Whitney, Elspeth, *Paradise Restored: The Mechanical Arts from Antiquity through the Thirteenth Century* (Philadelphia: American Philosophical Society, 1990).

Young, Mark Thomas, 'Nature as Spectacle: Experience and Empiricism in Early Modern Experimental Practice', *Centaurus*, 59. 1–2(2018), 72–96.

8   Teleology: A case study in iHPS

Andrea Gambarotto

Introduction

Teleology is a paradigmatic example of the benefits of integrating History and Philosophy of Science. The theoretical implications of teleology and its conceptual history are in fact often so intertwined that a historical perspective is absolutely essential to understand the philosophical issues at stake, let alone solve them.

Teleology is a kind of reasoning by which the goal of an entity contributes to the explanation of its very existence. In mainstream philosophy of biology, teleology is mostly considered a vestige of pre-Darwinian frameworks, such as natural theology, which appeals to a heavenly Designer whose intentions explain the existence of biological organisms. In this respect, J.B. S. Haldane (1892–1964), one of the fathers of modern population genetics, said that 'teleology is like a mistress to a biologist: he cannot live without her, but he is unwilling to be seen with her in public.'[1] This characterisation of teleology captures the fundamentally controversial character of the notion: on the one hand, teleology structures conceptualisations of biological systems; on the other hand, it continues to be treated as a residue of metaphysics that must be permanently eliminated from natural science.

The fundamental hypothesis of this chapter is that the controversial nature of teleology is essentially linked to a lack of appreciation of the complex semantics, both historical and theoretical, surrounding the concept. More precisely, my hypothesis is that Haldane's characterisation betrays a fundamental misunderstanding of teleology that can be traced back to the early modern period. This misunderstanding has hitherto prevented us from adequately conceptualising the intrinsic purpose of living systems.

By analysing some of the most relevant historical occurrences of teleology, this chapter aims to make a case for iHPS as a specific kind of inquiry, different from both externalist historical reconstruction and theoretical Philosophy of Science. I argue that a historical perspective is a crucial resource for the Philosophy of Science, because it serves to uncover the contextual origin of supposedly obvious concepts, bringing to light the assumptions on which they rest and the other theoretical alternatives that their emergence obscured.

This argument for iHPS, as Uljana Feest and Friedrich Steinle maintain, stems from a twofold theoretical commitment:

(1) on the one hand an interest in the historical processes by which scientific knowledge, as encapsulated in concepts, is generated, and (2) on the other hand a desire to construct philosophical accounts that do justice to the ways in which science is actually practiced (in this case: how scientific concepts are actually used or formed).[2]

At the same time, and this is my main point, a history-friendly attitude is not only useful (or necessary) for making sense of scientific concepts and investigative practices, but it also allows us to excavate archives of lost theoretical alternatives with the potential to open up conceptual solutions to contemporary issues in the Philosophy of Science. In this sense, this chapter should be read as an exercise in combining the History of Science and the History of Philosophy to come up with concepts that we can use to reframe our current theoretical problems from a different perspective.

As stressed by Ernst Mayr, 'the discussion of teleology occupies considerable space (10–14%) in several recent philosophies of biology'.[3] Yet, most philosophers have treated teleology as a unitary phenomenon, ignoring the fact that the term has been applied to several natural phenomena that are fundamentally different from one another. To appropriate an old Aristotelian expression, 'teleology is said in many ways'. Therefore, we must first of all understand what we are talking about when we use the term. In this respect, a historical perspective is fundamental for us to appreciate the concept's semantic stratifications. Indeed, by turning back to the term's Aristotelian roots, it is possible to identify a much different conception of teleology. With this goal in mind, my argument unfolds as follows:

Section 8.1 takes into account the early modern origins of the notion of teleology and its relation to Cartesian mechanism. I argue that this conception of teleology is fundamentally connected to an idea of purposiveness as (God's) intention and I underline some echoes of this strand of purposiveness in mainstream evolutionary theory. Section 8.2 turns back to the Aristotelian understanding of natural purposiveness, stressing its fundamental difference from the early modern conception of teleology and identifying when it reappeared after Kant. The reason I return to Aristotle is because of my general argument for integrating the history and philosophy of science; namely, that the history of science can be profitably understood not only as the 'sequence of past scientific theories,' but especially as an 'archive of theoretical alternatives'. Contained with this we can recover conceptual frameworks that can be applied to problems that still exist within philosophy of science today. Expanding upon this historical background, section 8.3 introduces a more detailed case study, concerning how the Aristotelian conception of teleology was implemented by Hegel. This will provide us with another notable example of how the history of science can provide useful conceptual tools to the philosophy of science. In fact, despite being largely regarded as purely metaphysical and unconcerned with natural science, Hegel's account of purposiveness represents the strongest re-emergence of the Aristotelian account of purposiveness as autonomous self-organisation. This view of teleology as autonomous self-organisation is slowly re-emerging in contemporary biology as well, and further demonstrates the importance of a discussion of teleology for issues within iHPS.

This chapter concludes by showing the recent resurgence of this idea of purposiveness in the work of the 'autopoiesis' school of Humberto Maturana and Francisco Varela. This will consolidate my claim that close attention to past theoretical endeavours can help reconceptualise natural purposiveness, by showing that a different understanding of teleology is possible and has indeed been put forth by a notable tradition, which can fruitfully be rediscovered today. My overall argument thereby suggests that an historical perspective can assist in more accurately framing contemporary problems with iHPS, identify interesting conceptual solutions, and help to avoid unfortunate cases of theoretical amnesia.[4]

8.1 Teleology and design: the early-modern conception of purposiveness and its place in contemporary biological theory

This section will focus on the early-modern understanding of natural purposiveness, which today still constitutes the main (and often sole) conceptual reference point for contemporary discussions of teleology. The term 'teleology' as we now use it was formulated by the philosopher Christian Wolff (1679–1754) in the first half of the eighteenth century. In his *Philosophia rationalis sive logica* (1728), Wolff defined 'teleology' as the part of physics that deals with the 'ends of things', describing it as a '*theologia experimentalis*' whose function was to make the wisdom of God comprehensible through the study of nature. This link between teleology and theology was built upon by the classic 'watchmaker analogy' of William Paley (1743–1805), according to which the organisation of living beings implies the existence of an 'intelligent designer' in the same way that the complex organisation of a watch implies the existence of a watchmaker. This premise is precisely what led contemporary philosopher of biology Michael Ruse to define teleology as a 'neo-scholastic pursuit'.[5] Our current mainstream concept of teleology is largely influenced by its early-modern origins and is grounded in the ideas of 'conscious intention' or 'intelligent design'. Since the early-modern period, this 'teleological' conception is opposed to a 'mechanical' conception of living organisms.

The explicit purpose of the usage of mechanistic conceptions of living organisms was to eliminate any reference to the final causes found in teleological conceptions. However, as Haldane pointed out, once thrown out the door, teleological language always seems to crawl back in through the window. Indeed, natural purposiveness is so structurally linked to the conceptual organisation of biological systems that mechanism seems unable to eliminate any of the teleological language that, albeit often inadvertent or disguised, remains at the heart of the discourse on living beings. Haldane's discourse is underwritten by a binary: either we explain the organisation of the living as the result of divine intention or we consider it the result of a pure mechanism lacking goal-directedness. However, this binary opposition is the result of a unilateral conception of teleology modelled solely on the kind of 'neo-scholastic' teleology described above. It is precisely here that a historical approach can help us put things into perspective, by understanding the origin of concepts which we otherwise take for granted.

Since Descartes, mechanism has been the way modern science conceives of scientifically legitimate explanation: as an opposition to teleological explanations. As a philosophical position, this implies two different arguments, which are often superimposed: (1) a metaphysical argument according to which living organisms are machines; (2) an epistemological argument according to which reference to efficient causes is the only means of legitimately explaining natural phenomena.

(1) In the *metaphysical* argument, mechanism as such is not incompatible with the idea that the order of the world is the result of the intention of God. In fact, though mechanical philosophers of the seventeenth century explicitly rejected final causes immanent to nature, they accepted an idea of purposiveness imposed on nature from the outside. In this sense, for Descartes or Newton, reference to efficient causes as the basis for scientific explanation was perfectly compatible with the fact that those causes were the product of a divine plan.

(2) In the *epistemological* argument, the idea of immanent final causes continued to play an important, though often not explicitly recognised, role in the natural philosophy of the 17th century. Although this period is generally understood as the heyday of mechanism, studies have revealed the presence of teleological analyses in representative natural philosophers such as Robert Boyle or Pierre Gassendi,[6] and even a 'more discrete presence of functional explanatory concepts in self-proclaimed mechanists such as Descartes'[7] which has recently led the philosopher and historian of science Charles T. Wolfe to label the methodology of modern natural philosophy with regard to the explanation of living organisms as 'teleomechanism'.

If mechanism is essentially a reaction against teleological arguments, Darwinism is generally considered its battle horse. In Darwinism, even though organisms appear to be the result of design, there is no designer: natural selection alone performs this role. Considered from a historical perspective, however, this approach also maintains some forms of teleological language. In the context of contemporary evolutionary biology, the divine architect is replaced by natural selection, a purely mechanical process that is considered sufficient to eliminate all teleology from scientific discourse. Yet, as Michael Ruse says, 'the metaphor of design, with the organism as an artefact, is at the heart of Darwinian evolutionary biology'.[8] This statement is confirmed by studies in different fields. Historian of science Dov Ospovat (1981) has shown that Paley's natural theology played a key role in Darwin's scientific training in Cambridge, before he left on the Beagle. More recently, philosopher Tim Lewens (2006) has pointed out that, from a conceptual point of view, the dominant language in contemporary evolutionary biology is strikingly similar to William Paley's natural theology. In a similar way, biologist John Reiss (2009, p. xiii) argues that:

> the relation of current evolutionary biology to the 'design problem' is indeed rather strange. We deny evolution any teleology, any goal-directedness. Yet we use past selection as a way to explain the current adaptedness of organisms, in much the same way that pre-Darwinian natural theologians invoked the past actions of the Creator – as exemplified by Paley's famous metaphor of the Creator as divine watchmaker.

This artefactual (and therefore somewhat theological) conception of biological organisation was also the critical target of the now classic article by Stephen Jay Gould and Richard Lewontin (1979) entitled 'The Spandrels of San Marco and the Panglossian Paradigm: A Critique of the Adaptationist Programme'. The 'Panglossian Paradigm' refers here to the satire *Candide* by Voltaire, in which the French philosopher criticises a caricatured version of Leibnizian philosophy in which 'everything that happens is in the best of worlds wanted by God'. In a rather provocative way, Gould and Lewontin consider this mantra as also the underlying premise of Orthodox Darwinism's 'adaptationist' programme: if a trait was selected, then it must be the 'optimal' one.

Gould and Lewontin's argument is probably the best example of what I have argued in the present section, namely that our mainstream understanding of purposiveness in nature stems largely from the early-modern conception of teleology as intention. In the following section, I will argue that historical examples show us that the artefactual paradigm is not the only possible way to conceptualise the teleological features of biological organisation. In fact, historically speaking, this conception of teleology derives from a different understanding of natural purposiveness, one strongly underrepresented in current mainstream debates: the rich and yet largely neglected tradition, in philosophy as well as in biology, of an Aristotelian conception of finality as the intrinsic feature of living systems.

8.2 A minor historical canon: intrinsic teleology and its tradition

In the previous section I showed how the early-modern conception of teleology as 'intelligent design' still plays a fundamental role in our mainstream neo-Darwinian conception of natural purposiveness. I turn now to the older, Aristotelian understanding of organic purposiveness, in order to stress that the early-modern understanding of teleology sketched above is not the only possible conceptualisation. I will then briefly sketch the resurgence of Aristotelian natural purposiveness after Kant in the nineteenth and twentieth centuries when teleology was not understood in terms of the assumption (or metaphor) of conscious intention but rather in terms of autonomous self-organisation.

Aristotle never used a Greek equivalent for the term 'teleology', he instead mainly referred to 'the cause for the sake of which'. In almost every passage in which Aristotle introduces, discusses, or argues for the existence of final causality, his attention is focused on the generation and development of living organisms. What does Aristotle mean when he says that development is *for the sake of* the resulting adult organism? In an article that has become classic on the subject, Allain Gotthelf argues that most existing interpretations of Aristotle's words fit into two main traditions:

1 The first, which may be called the 'immaterial agency' interpretation, maintains that Aristotle understands natural teleology fundamentally on an analogy to human purposive action, in a way that implies that the developing embryo is (or embodies) some sort of conscious or quasi-conscious agent, directing the flow of materials, guiding its development to maturity.

2 The second, and more recent interpretation, may be designated the 'explanatory condition' interpretation. It maintains that the 'final cause' is in no sense actually a cause, though it plays a role in a certain type of explanation; and that its role in the explanation of organic development is to identify that for which the stages of development are necessary and (thus) that in terms of which they must be identified if they (and the development they constitute) are to be rendered 'intelligible.'[9]

According to Gotthelf, both interpretations lead to 'Aristotelian mortal sins': the first reverses the Aristotelian priority of nature over art; the second seems to suggest that mechanical, and thus fortuitous events, are 'for the sake of something'. Gotthelf concludes that Aristotelian teleology is 'neither vitalist and mystical, nor "as if" and mechanical'.[10] On the contrary, for Aristotle purposiveness is an entirely natural, constitutive and intrinsic, feature of living beings. In this sense, the philosopher Marjorie Grene (1972, p. 398 cited in Mayr 1992, p. 121) is right to point out that 'there is absolutely no question of "purpose" here, either man or God's. To suppose otherwise is to introduce a Judeo-Christian confusion of which Aristotle must be entirely acquitted'.

The historical re-emergence of this idea of teleology in modern times is connected to the work of Immanuel Kant. In the first paragraphs of the 'Critique of the Teleological Judgment', the second section of his *Critique of the Power of Judgment* (1790), Kant introduces a fundamental distinction between 'extrinsic purposiveness' and 'intrinsic purposiveness'. This distinction is essentially intended to mark the difference between artefacts and organisms. Extrinsic purposiveness can be defined as 'utility' and is a constitutive feature of artefacts: for instance, the purpose of a watch is to mark the hours. It is defined as 'extrinsic' in these cases because purposiveness is never a property of the artefact as such but always extrinsically attributed to it by a designer. On the other hand, intrinsic purposiveness defines the fundamental phenomena of living beings: reproduction, growth and functional integration. These are defined as 'intrinsic' because they are not produced by some outside being but rather by the living organism itself.

As an example of intrinsic purposiveness, Kant takes the case of a tree. A tree produces itself in relation to the species, the individual, and its parts: it generates other trees of the same species; it generates itself as an individual through the phenomenon called growth; and this generation works in such a way that the preservation of each part is mutually dependent on the preservation of the others. Organised beings that display these characteristics are defined by Kant as 'natural purposes' (*Naturzwecke*). Natural purposes have an 'intrinsic' purpose, that is, self-organising features that are absent in machines. In a watch, in fact, each part is organised in relation to the others, but the watch itself does not produce them. On the basis of these considerations, Kant asserts that:

> an organized being is thus not a mere machine, for that has only a *motive* force, while the organized being possesses in itself a *formative* force (*Bildungskraft*), and indeed one that it communicates to the matter, which does not have it (it organizes the latter): thus it has a self-propagating formative power, which cannot be explained through the capacity for movement alone (that is, mechanism).[11]

The development of such an intrinsic conception of teleology is linked to the empirical work of certain post-Kantian naturalists belonging to the so-called 'Göttingen School' as well as to the enterprise of Romantic *Naturphilosophie*. At the beginning of the nineteenth century, discussing this conception of teleology, Hegel said that 'one of the great services rendered by Kant to philosophy consists in the distinction he has established between the relative or extrinsic finality and the intrinsic finality; in the latter, he opened the concept of life'.[12]

This very brief sketch of the 'intrinsic teleology tradition' that I am summarily reconstructing in this section goes all the way into the twentieth century. In the early twentieth century, a number of theoretical biologists, such as J.S. Haldane, D'Arcy W. Thompson, E.S. Russell, J.H. Woodger, J. Needham, and C.H. Waddington in England, P.A. Weiss and L. von Bertalanffy in Austria, and J. von Uexküll in Germany developed this idea of teleology into a new non-reductionist, or 'organicist', theory of the organism.[13] Due to its theoretical continuity with the nineteenth-century German tradition's emphasis on organic wholes and intrinsic teleology, this twentieth-century intervention was recently labelled, using a theoretically productive anachronism, as 'Romantic biology'.[14]

This approach had some currency in the first half of the twentieth century, but was completely obliterated by the molecular revolution starting in the 1940s and has since then been largely underrepresented on the contemporary biological scene.

As philosopher Denis Walsh (2006, p. 774) puts it, 'contemporary biology has responded to the problem posed by the natural purposiveness of organisms by the simple expedient of ignoring it.' However, even if Kant's problem has been largely forgotten in biology, it strongly resonates with issues that are only now beginning to attract biologists' attentions – self-organization, the 'emergent' properties of organisms, their adaptability, their capacity to regulate their component parts and processes.[15] Biology is currently witnessing an important transition toward less 'reductionist' approaches to life that favour an 'organicist' view of living systems.[16] A central implication of this transition is reappraisal of the idea of 'organism' that was displaced in the 1940s by the establishment of modern evolutionary synthesis and the flourishing of molecular biology.[17]

In the context of this shift, reference to frameworks and concepts elaborated by authors belonging to the post-Kantian tradition is strikingly frequent: Stephen Jay Gould and Richard Lewontin, for instance, have mobilised the concept of *Bauplan* in order to criticise the adaptationist programme of modern evolutionary synthesis;[18] Scott Gilbert and Sahotra Sarkar (2000) have argued that Kant's view of organisms was a fundamental precondition of developmental biology; and Stuart Kauffman (2000) has even portrayed Kant's understanding of organised beings as a template for self-organisation theory.

In this sense, as historian Georg Toepfer (2012) puts it, the most recent developments in contemporary biology overthrow Theodosius Dobzhansky's famous motto, 'nothing in biology makes sense except in the light of evolution' and lead us to the idea that 'nothing in biology makes sense except in the light of teleology'. This situation justifies looking back at the tradition I have been sketching not only with the scholarly eye of the pure historian but in search of theoretical frameworks that can be applied to our contemporary problems. In this section I have presented only a very summary sketch of a century-long tradition conceiving of teleology in terms of autonomous self-organisation, as opposed to the design framework that characterises the early-modern origins of the term and its use in contemporary biological theory. In the following section I implement this general background with a specific case study, namely Hegel's philosophical account of teleology, sketching a few remarks about how this work can help in trying to define a general methodology for iHPS.

8.3 Teleology without intention: Hegel's philosophical account of purposiveness

This section extends the previous sketch by analysing Hegel's philosophical account of teleology. It thereby provides an example of how history of science and history of philosophy can be productively integrated to approach the same problem from complementary perspectives. On this basis, in the concluding remarks to this section I will provide a brief sketch of a possible methodology for iHPS. In a nutshell, this methodology consists, on the one hand of mobilising the language of contemporary Philosophy of Science to 'translate' older concepts that would remain obscure if left to the specific jargon of their time. This 'translation', however, is not aimed at arguing that those historical categories simply 'anticipated' some of our contemporary views. Rather, I argue that the focus of iHPS should precisely be the resistance of some historical categories to our contemporary frameworks, for precisely that resistance indexes their possibility of bringing to light philosophical assumptions that are not discussed in contemporary debates.

Unlike his peers, Kant and Schelling, who contributed concretely to the formulation of empirical research programmes for emerging biological science, Hegel did not actively participate in scientific controversies. Although he had been well aware of the developments occurring in most scientific disciplines of his time since his early years in Jena, his overall attitude was one of external observation and comment. In this sense, it is worth mentioning that while Hegel was composing his *Jaener Entwürfe* in the early years of the nineteenth century, the epistemological space for the emergence of biology as a unified field had already been opened and the process of its institutionalisation had already begun. Nevertheless, or perhaps precisely because of this non-involvement, Hegel was able to criticise both Kant and the Romantic *Naturphilosophie*, as well as to offer one of the most eloquent accounts of the main philosophical points at stake in this debate.

Hegel criticised the Kantian account of purposiveness with vehemence in the 'Teleology' section of the *Science of Logic* and the *Encyclopedia*. This is the section of his System in which Hegel takes into account the conceptual structure of 'extrinsic' purposiveness, i.e. utility. Here, Hegel discusses the logical structure of the final causes proper to instrumental or technical entities, whose essence lies in the fact of their being *for* something. 'Intrinsic' teleology, in Kant's sense, as the specific purposiveness displayed by organised beings, is discussed in the following section, which in fact is entitled 'Life'. The idea of life as already containing inner purposiveness stands for Hegel 'infinitely far beyond the concept of modern teleology which has only the *finite*, the *extrinsic* purposiveness in view',[19] since here 'there is the perception of *purposiveness*, an *intelligence* is assumed as its author'.[20] However, in Hegel's view, the concept of an extra-ordinary intelligence is proximate to the teleological principle, and thus seems to depart from the true investigation of nature.

With his typical philosophical methodology, Hegel intended to disclose the conceptual potentialities implicit in the notion of purpose and to show that technical purposiveness, i.e. *extrinsic* purposiveness, is a derivative form of teleology, presupposing a more fundamental kind of purposiveness, namely *intrinsic* purposiveness. Hegel argues that whenever the teleological relation is understood as extrinsic purposiveness, organisation is not conceived as the result of processes internal to the living individual itself but rather as the result of an external agency.

Kant criticised Gottfried Wilhelm Leibniz's understanding of artifacts as *human* machines and of living organisms as *divine* machines. For Leibniz, the only difference between the two categories is that the latter are *infinitely* organised, unlike the former. This conceptual distinction is a quantitative matter; it does not imply a qualitative, categorical difference between machines and organisms. Hegel upholds Kant's rejection of this idea as his predecessor articulated it in his *Critique of the Power of Judgment*, but at the same time argues that Kant was not able to consistently move beyond this position either. Kant's controversial resolution of the antinomy of teleological judgment is at least partially the result of his middle-position on the argument about intelligent design.

As Hegel points out, on the one hand, Kant criticised modern metaphysics for understanding natural purposiveness in terms of mere utility and emphasised the incongruence of referring to a divine maker when talking about natural objects. His distinction between 'internal' and 'external' purposiveness is aimed precisely at discarding this metaphysical assumption. On the other hand, Kant is quite firm in maintaining that we can only conceive of natural organisation according to the model of artefactual organisation. He therefore construes purposiveness as a regulative principle: we cannot claim that organised beings are the result of divine design, but we do not seem to have other options for conceiving of them.

Thus, according to Kant (2000, p. 271) we must consider organised beings *as if* they were the result of intention, while looking for mechanical explanations of biological organisation:

> we can boldly say that it would be absurd for humans even to make such an attempt or to hope that there may yet arise a Newton who could make comprehensible even the generation of a blade of grass according to natural laws that no design (*Absicht*) has ordered; rather, we must absolutely deny this insight to human beings.

Indeed, according to Kant, biological organisation cannot be explained without reference to an intention, but this reference is unsustainable on empirical grounds alone: it must be understood as a heuristic device. Thus, despite his distinction between types of purposiveness, Kant like Leibniz still conceives purposiveness as always related to an intention.

Hegel argues that when we consider living organisms in this way, we do not really understand them as purposes. Instead they are merely means that are used to realise some purpose lying outside of them. According to this framework, a living organism is not considered a natural purpose *per se*; its purposiveness is rather understood as the result of an external agency, and in this sense, there is no substantial difference between organisms and artefacts, between a living being and a clock. On the other hand, Hegel's aim is precisely to comprehend the living organism '*by itself* as a *whole*', a totality 'that is not found in purpose and the extra-mundane intelligence associated with it'.[21] For Hegel, conceiving purposiveness as *extrinsic* always implies a conscious intelligence that previously defined the object under consideration through reference to an idea that exists in and for itself – an idea which is not the result of a process of *self*-organisation.

In Hegel's view such external teleology underlies the conceptual structure of artefactual entities, whose purpose is not identical to them but something external. In this case, the content of a concept, which for Hegel is the specific form of organisation to a living individual, is externally given rather than autonomously produced. On this score, Hegel (1981, p. 183) maintains that 'one of Kant's greatest services to philosophy was in drawing the distinction between relative or *extrinsic* purposiveness and *intrinsic* purposiveness; in the latter he opened up the concept of *life*, the *idea*'. At the same time, however, Hegel felt that Kant did not follow through with this insight, ultimately abiding by an extrinsic, intentional conception of teleology. This censure of Kant's treatment of teleological judgments 'remains as constant in Hegel's writings as ... his praise of the idea of inner purposiveness'.[22]

For Hegel an adequate understanding of the notion of internal purposiveness is what opens up the idea of life. He defines life as that phenomenon which is a purpose in and for itself. Something is alive when there is no opposition between purpose and organisation, because the organisation is both the cause and effect of itself. Since the organisation of a natural purpose is not the result of some external design but internally produced, it implies a form of self-determination, which Hegel calls *subjectivity*. This form of self-determination is the hallmark of autonomy: 'this *subject* is the idea in the form of singularity, as simple but negative self-identity – the *living individual*'. This individual, or subject, is characterised by Hegel as 'the initiating self-moving *principle*'.[23]

Within this framework, because of its self-determination, 'the *purposiveness* of the living being must be grasped as *intrinsic*' and must be understood as the living being's most essential characteristic, which distinguishes it from externality and pervades it thoroughly: 'this objectivity of the living being is the *organism*; it is *means* and *instrument* of purpose, fully purposive.' For this reason, the organism is a manifold not of *parts* but of *members*: when separated from it, 'they revert to the mechanical and chemical relations common to objectivity' This idea of inner purposiveness is for Hegel 'the *concept of the living subject* and of its *process*', whose hallmark is a form of 'self-referring negative unity.'[24]

In other words, what makes something *living* is precisely the peculiar organisation of its parts: this organisation is not something static but a dynamic process in which every part contributes to the subsistence of the whole, and thus the whole can be said to be at the same time the cause and the effect of itself. This form of relation to oneself (which in Hegelian terms is subjectivity) constitutes the fundamental structure of life and is described according to three different levels of autonomy: (1) the *living individual* (*das lebendige Individuum*), (2) the *living process* (*Lebensprozess*), and (3) the *species* (*Gattung*). These three determinations constitute the core of Hegel's theory of biological individuality, a theory that is expounded in the *Science of Logic*.

In that text, Hegel discusses (1) the living individual by means of close conceptual analysis of the form of organisation that characterises biological individuals. Here Hegel makes reference to the 'vital forces'. The notions of *sensibility, irritability* and *reproduction* are transfigured to fit the theoretical framework of Hegel's logic and made to correspond to the three logical instances of *universality, particularity* and *singularity*. Hegel depicts sensibility as the 'external existence of the inward soul', which reduces externality to the 'complete simplicity of the self-equal universality'; irritability is instead defined as the 'vital force of resistance (*Widerstandkraft*)' by means of which the living individual reacts to its surrounding environment. It is only in reproduction, however, that 'life is *something concrete* and vital; in it alone does it also have feeling and power of resistance'.[25] In this context, repro-duction should be understood not as the reproduction of the *species* but rather as the reproductive process internal to the *individual* itself: its faculty of continuously regenerating its own members. Hegel goes on to account for (2) the living process as the dialectical relation of an organised individual to the world external to it. This process begins with *need* (*Bedürfnis*), which is defined by Hegel as a form of 'self-determination of the living being'. Some-what paradoxically, the identity of an organism is marked by its continuous exchange with the environment. Through the experience of need, the living subject is connected with the outside world and seizes hold of the objects it finds there by means of the metabolic process.

Finally, Hegel characterises (3) the species (*Gattung*) as 'the completion of the idea of life', because in the species a living individual displays an 'identity of itself with its hitherto indifferent otherness'. This means that the living individual is at the same time identical to and different from the rest of its species. *Identical* since they belong to the same universal set, *different* because that set is made up of particular individuals who are *ipso facto* different from each other. In other words, 'externality is the individual's immanent moment and is, moreover, itself a living totality; an externality in which the individual has certainty of itself not as *being sublated*, but as *subsisting*.'[26] For this reason, the defining characteristic of living individuals is their impulse to self-replicate: 'from this side the species obtains *actuality* through its reflection into itself, for the moment of negative unity and individuality is thereby *posited* in it – the *propagation* of the species.'[27]

Through analysis of these three determinations – the individual, the living process and the species – the *Science of Logic* offers the conceptual basis for Hegel's treatment of biological teleology. Here Hegel provides a conceptual framework that describes the essential characteristics of the living: organisa-tion, need, and self-replication. Through these ideas, Hegel provides the out-line for his theory of living subjectivity, which is defined as the form of being-for-self that realises itself in relation to its otherness. The most important result of his philosophical analysis is the observation that Kant's account of teleology is underwritten with the unspoken assumption that purposiveness can only be understood as analogous to conscious intention. To the contrary, in accordance with the rest of the post-Kantian tradition, Hegel interprets teleology in terms of self-organisation and autonomy.

Conclusion: applications for current iHPS

As a conclusion to this analysis, I would like to turn back to my general argument concerning the integration of historical and conceptual approaches, providing a brief sketch of how 'Hegel's philosophy of biology' could be approached from an iHPS perspective. This methodology would imply, on the one hand, a historical reconstruction of Hegel's scientific context and a detailed analysis of the logic inherent to Hegel's philosophical treatment of the scientific issues of his time. In other words, it would analyse the way in which Hegel worked with existing scientific concepts to construct his theory of organic nature. At the same time, its goal would be to show how this theory relates to contemporary debates within the philosophy of biology. Although these might appear to be two mutually exclusive objectives, the combination of historical and theoretical perspectives yields an innovative, and notably more comprehensive understanding of Hegel's philosophy of biology.

Importantly, this work would not aim to demonstrate that Hegel 'anticipated' some of our contemporary views on the nature of living systems. Rather, it would mobilise the language of contemporary philosophy of biology to 'translate' Hegelian concepts that would remain obscure if left in the language of Romantic philosophy of nature. Accordingly, it would also stress the resistance of Hegelian texts to our contemporary categories, for that resistance indexes their possibility of bringing to light philosophical assumptions that are not discussed in contemporary debates. Every time we meet a form of such 'resistance', we need to ask whether that resistance marks a contingency, i.e., the fact that Hegel was dealing with theories that have since been proven wrong and thus cannot be translated into our contemporary language, or if the resistance is an opportunity for us to think about biological problems in a fresh and original way. For instance, the fact that Hegel uses the concept of 'subjectivity' to discuss animal organisms is not necessarily historically contingent. Rather, it might be a consequence of the fact that Hegel proposed a definition of subjectivity that is absent from our contemporary biological vocabulary. In fact, Hegel's understanding of biological organisms as living 'subjects' is perhaps one of the most innovative aspects of his philosophy of biology – one which might intersect and corroborate various non-mainstream approaches in contemporary biological thought, including the theories of autopoiesis,[28] dialectical biology,[29] self-organisation,[30] and biological autonomy.[31]

To give a concrete example of what I have been describing above, I would now briefly like to show how this Aristotelian/Hegelian idea of purposiveness is currently witnessing a renaissance in some non-mainstream approaches to philosophy of biology. I thereby turn back to the general point I made in the introduction: that a historical perspective can bring to light an archive of conceptual solutions able to fuel innovative perspectives on our contemporary issues in Philosophy of Science.

In his very last paper, 'Life after Kant', Francisco Varela argued for the

> great need to bring to the fore the remarkable and recent convergence between the re-awakening of the philosophical discussion concerning natural purposes (with Jonas as the central figure), and an independent but convergent stream of thought concerning biological individuality and the organism (with the autopoiesis school as the central figure).[32]

In virtue of these developments, Varela deems it possible that 'after two centuries, we can move *beyond* the unstable position set out by Kant in the *Critique of Judgment*, and therefore provide a fresh re-understanding of natural purpose and living individuality.'[33] Indeed, Kant held an unstable position by arguing *on the one hand* for the impossibility of a reductionist account of organisms, while *on the other hand* maintaining that the teleological features displayed by living systems should only be considered heuristically, not as part of their ontological character.

The received view of teleology in contemporary, mainstream philosophy of biology puts the emphasis especially on the latter aspect, reading Kant as a forerunner of current 'artifact models' of the organism, according to which biological purposiveness should be understood as the result of a statistical, optimizing process carried out by evolution through natural selection. On the other hand, as Hegel before him, Varela insists that if we want to 'take teleology seriously' as a natural phenomenon, we need to fully develop the former aspect of Kant's controversial legacy.

If we consider my previous analysis, Varela's proximity to Hegel is quite striking: Hegel likewise recognised Kant's understanding of organised beings as natural purposes but criticised him for conceiving of this purposiveness merely in terms of intentional agency and ascribing it a merely regulative function.

The proximity between Varela and Hegel is even more evident when Varela writes that:

> teleology, understood as intrinsic teleology, turns out to be an empirical feature of an organism, its *sine qua non* condition. But this is objective not in an absolute sense, only insofar as an organism is a center that organizes matter into a living being and its *Umwelt*, hence enacting on this stage the original split of subject and its world and their dialectical interrelatedness.[34]

The manifestly Hegelian flavour of this last quote is a useful conclusion to this paper and provides one last piece of evidence to its central claim: that a historical perspective is a crucial resource for Philosophy of Science – one which can help us uncover the contextual origin of supposedly obvious concepts, frame our philosophical problems correctly, and provide a rich archive of potential solutions.

Notes

1 Cited by Ernst Mayr 1974), p. 115.
2 Feest and Steinle, (2012), p. 8.
3 Ernst Mayr (1992), p. 117.
4 Sections 8.1 and 8.2 are largely the results of my discussions with Matteo Mossio, section 8.3 contains material that has already appeared in my book, *Vital Forces, Teleology and Organization: Philosophy of Nature and the Rise of Biology in Germany* (Gambarotto, 2018).
5 Michael Ruse, 'The last word on teleology, or optimality models vindicated', in Ruse (1981), p. 85.
6 James G. Lennox (1983); Margaret J. Osler (2001).
7 Charles T. Wolfe (2014), p. 292. Cf. also Stephen Gaukroger (2000)p. 383–400.
8 Michael Ruse (2003), p. 266.
9 Gotthelf 1976, pp. 252–253.
10 Gotthelf 1976, p. 253.
11 Immanuel Kant, *Kritik der Urteilskraft* (Akademie Ausgabe, 1968), Band V, English translation by Paul Guyer and Eric Matthews, Kant (2000), p. 246.
12 Georg W.F. Hegel (1981), p. 241, English translation by George Di Giovanni, *The Science of Logic* (Cambridge, Cambridge University Press, 2010).
13 Donna J. Haraway (1976); Erik L. Peterson (2017); Daniel J. Nicholson and Richard Gawne (2015).
14 Maurizio Esposito (2013).
15 Walsh 2006, p. 772.
16 Massimo Pigliucci and Gerd Müller (2010); Kevin Laland, et al., (2014); Philippe Huneman and Denis Walsh (2017).
17 Philippe Huneman and Charles T. Wolfe (2010).
18 Gould and Lewontin (1979).
19 Georg W.F. Hegel (1992), § 204.
20 Hegel (1981), p. 651.
21 Hegel (1981), pp. 155–156.
22 Daniel Dahlstrom (1998), p. 168. Cf. also Franco Chiereghin (1990).
23 Hegel (1981), p. 183.
24 Ibid., pp. 184–185.
25 Ibid., p. 186.
26 Ibid., p. 190.
27 Ibid., p. 191.
28 Maturana and Varela, (1980).
29 Levins and Lewontin (1985).
30 Kauffman (2000).
31 Moreno and Mossio (2015).
32 Weber and Varela (2002), pp. 97–98.
33 Matteo Mossio and Leonardo Bich (2014), p. 1090.
34 Weber and Varela (2002), p. 120.

Bibliography

Chiereghin, F., 'Finalità e idea della vita: La ricezione hegeliana della teleologia di Kant', *Verifiche*, 19. 1(1990), 127–230.

Dahlstrom, D., 'Hegel's Appropriation of Kant's Account of Teleology in Nature', in *Hegel and the Philosophy of Nature*, ed. by S. Houlgate (New York: Suny Press, 1998), pp. 167–188.

Esposito, M., *Romantic Biology, 1890–1945* (Abingdon: Routledge, 2013).

Feest, U., and Steinle, F., *Scientific Concepts and Investigative Practice* (Berlin: de Gruyter, 2012).

Gambarotto, A., *Vital Forces, Teleology and Organization: Philosophy of Nature and the Rise of Biology in Germany* (Dordrecht: Springer 2018).

Gaukroger, S., 'The Resources of a Mechanist Physiology and the Problem of Goal-Directed Processes', in *Descartes' Natural Philosophy*, ed. by S. Gaukroger, J. Sutton, et al. (Abingdon: Routledge, 2000), pp. 383–400.

Gilbert, F.S., and S. Sarkar, 'Embracing Complexity: Organicism for the 21st Century', *Developmental Dynamics*, 209(2000), 1–9.

Gotthelf, A., 'Aristotle's Conception of Final Causality', *Review of Metaphysics*, 30. 2 (1976), 226–254.

Gould, S.J., and Lewontin, R.C., 'The Spandrels of San Marco and the Panglossian Paradigm: A Critique of the Adaptationist Programme', *Proceedings of the Royal Society of London*, 1191(1979), 581–598.

Grene, M., 'Aristotle and Modern Biology', *Journal of the History of Ideas*, 33. 3 (1972), 395–424.

Haraway, D.J., *Crystals, Fabrics and Fields: Metaphors in Twentieth-Century Developmental Biology* (New Haven: Yale University Press, 1976).

Hegel, G.W.F., *Wissenschaft der Logik, zweiter Band, Die Subjektive Logik*, ed. by F. Hogemann and W. Jaeschke (Hamburg: Felix Meiner, 1981).

Hegel, G.W.F., *Enzyklopädie der Philosophischen Wissenschaften im Grundrisse (1830)*, ed. by W. Bonsiepen and R. Heede (Hamburg: Felix Meiner, 1992).

Huneman, P., and Walsh, D., *Challenging the Modern Synthesis: Adaptation, Development and Inheritance* (Oxford: Oxford University Press, 2017).

Huneman, P., and C. Wolfe, 'The Concept of Organism: Historical, Philosophical, Scientific Perspectives', *History and Philosophy of the Life Sciences: Special Issue*, 32. 2–3(2010).

Kant, I., *Critique of the Power of Judgment*, translated by P. Guyer and E. Matthews (Cambridge: Cambridge University Press, 2000).

Kauffman, S., *Investigations* (Oxford: Oxford University Press, 2000).

Laland, K.N., T. Uller, M.W. Feldman, and others, 'Does Evolutionary Theory Need a Rethink?', *Nature*, 514(2014), 161–164.

Lennox, J.G., 'Boyle's Defense of Teleological Inference in Experimental Science', *Isis*, 74. 1(1983), 38–52.

Levins, R., and R. Lewontin, *The Dialectical Biologist* (Cambridge MA: Harvard University Press, 1985).

Lewens, T., *Organisms and Artifacts: Design in Nature and Elsewhere* (Cambridge, MA: MIT Press, 2006).

Maturana, H., and Varela, F., *Autopoiesis and Cognition: The Realization of the Living* (Dordrecht: Springer, 1980).

Mayr, E., 'Teleological and Teleonomic: A New Analysis', in *Methodological and Historical Essays in the Natural and Social Sciences*, Boston Studies in Philosophy of Science, (Dordrecht: Reidel, 1974), pp. 91–118.

Mayr, E., 'The Idea of Teleology', *Journal of the History of Ideas*, 53. 1(1992), 117–135.

Moreno, A., and M. Mossio, *Biological Autonomy: A Philosophical and Theoretical Enquiry* (Dordrecht: Springer, 2015).

Mossio, M., and L. Bich, 'What Makes Biological Organization Teleological', *Synthese*, 194(2014), 1089–1114.

Nicholson, D.J. and R. Gawne, 'Neither Logical Empiricism, Nor Vitalism, But Organicism: What Philosophy of Biology Was', *History and Philosophy of the Life Sciences*, 37. 4(2015), 345–381.

Osler, M.J., 'Whose Ends? Teleology in Early Modern Natural Philosophy', *Osiris*, 16 (2001), 151–168.

Ospovat, D., *The Development of Darwin's Theory: Natural History, Natural Theology, and Natural Selection, 1838–1859* (Cambridge: Cambridge University Press, 1981).

Peterson, E.L., *The Life Organic: The Theoretical Biology Club and the Roots of Epigenetics* (Pittsburgh: University of Pittsburgh Press, 2017).

Pigliucci, M., and G. Müller, *Evolution: The Extended Synthesis* (Cambridge, MA: MIT Press, 2010).

Reiss, J., *Not by Design: Retiring Darwin's Watchmaker* (Berkeley: University of California Press, 2011).

Ruse, M., *Is Science Sexist? And other Problems in the Biomedical Sciences* (Dordrecht: Reidel, 1981).

Ruse, M., *Darwin and Design: Does Evolution Have a Purpose?* (Cambridge, MA: Harvard University Press, 2003).

Toepfer, G., 'Teleology and its Constitutive Role for Biology as the Science of Organized Systems in Nature', *Studies in History and Philosophy of Biological and Biomedical Sciences*, 43. 1(2012), 113–119.

Walsh, D., 'Organisms as Natural Purposes: The Contemporary Evolutionary Perspective', *Studies in History and Philosophy of Biological and Biomedical Sciences*, 37. 4(2006), 771–791.

Weber, A., and F. Varela, 'Life after Kant: Natural Purposes and the Autopoietic Foundations of Biological Individuality', *Phenomenology and the Cognitive Sciences*, 97. 1(2002), 97–125.

Wolfe, C.T., 'Teleomechanism Redux? Functional Physiology and Hybrid Models of Life in Early Modern Natural Philosophy', *Gesnerus*, 71. 2(2014), 290–307.

9 The cybernetic origins of enactivism and computationalism

Joe Dewhurst

Introduction

The enactivist tradition in philosophy of mind and cognitive science is often seen as standing in opposition to the mainstream computationalist tradition, despite the shared ancestry of the two traditions in mid-twentieth century cybernetics.[1] This chapter traces the development of enactivist thought from its cybernetic origins, via the autopoietic theory of the Chilean cyberneticist Humberto Maturana. The purpose of this analysis is twofold: firstly, to illustrate that many of the ideas central to enactivism need not necessarily entail an opposition to computational characterisations of cognition; and secondly, to identify exactly how and why enactivism came to be seen as an anti-computationalist tradition. An iHPS approach is crucial for both of these points, as without understanding the historical roots of contemporary cognitive science it is difficult to see why these two traditions became opposed in the first place, or how we might be able to move forward towards a future reconciliation.

Section 9.1 introduces enactivism and computationalism, and explains why these traditions are usually seen as being opposed to one another.[2] Section 9.2 describes the respective origins of the two traditions in the cybernetic notions of biological homeostasis and neural computation, and looks briefly at how these two notions were, for a time, able to co-exist. Section 9.3 traces the development of the enactivist notion of autonomy, from its roots in biological homeostasis, through Maturana's autopoietic theory,[3] to the origins of modern enactivism in Varela, Thompson, and Rosch's *The Embodied Mind*. Finally, section 9.4 returns to the contemporary dispute between enactivism and computationalism, and argues that the enactivist notion of autonomy is only incompatible with computationalism if computation is understood as a semantic phenomenon. By adopting a non-semantic, mechanistic account of computation, one that in fact has much in common with cybernetics and autopoietic theory, it might be possible to reconcile these two traditions.

9.1 Enactivism and computationalism

Contemporary cognitive science is dominated by theories and approaches that treat the brain (and nervous system) as a computational system of some kind.[4] These approaches have been experimentally productive and have enjoyed apparent empirical success.[5] They are supported philosophically by the computational theory of mind, or computationalism for short.[6] Computationalism holds that mental states and processes are the product of computations of some kind, and appeals to the success of computational cognitive science in order to explain how the mind works. There is a long history of opposition to the computational theory of mind, but perhaps its greatest opponent can be found in the various enactivist traditions, which typically hold as a core tenet that cognition cannot be computational.[7] This section will briefly review the basic details of both computationalism and enactivism, before outlining the reasons enactivism gives for rejecting computationalism.

9.1.1 Computationalism

The computational theory of mind developed out of work associated with cybernetics, but it was given its first proper philosophical articulation by the American philosopher Hilary Putnam in the 1960s.[8] Putnam (1967) argued for a view that has since become known as 'machine functionalism', according to which mental states and processes should be understood as certain kinds of functional organisation, which could in principle be implemented by a correctly organised computational system. Putnam's student Jerry Fodor (1975) subsequently developed this general approach into a more specific computational theory of mind, according to which mental states are composed of language-like symbols constituting a 'language of thought'. According to Fodor, the language of thought is meant to provide a computational implementation for the mental state attributions of common sense folk psychology, whose operations and transitions coordinate the behaviour of the system within which it is implemented. To give a toy example, a perceptual stimulus might generate the symbol COW, which then interacts with the symbols BURGER and HUNGER, and causes the system to engage in hunting behaviour. It is (relatively) easy to imagine how a system of this kind might be implemented computationally (however difficult it might be in practice), thus offering a plausible naturalisation of our folk theory of mind.

The idea that the brain is a computer is heavily embedded in contemporary cognitive science, and has been ever since the discipline originally emerged out of the cybernetics movement. The field of computational neuroscience, for example, has been relatively successful at uncovering the functional structure of the brain, even if it is still unclear how this structure relates to 'thinking' as such.[9] More generally, computational methods and models permeate the cognitive sciences, although the extent to which these actually require us to think of the brain itself as computational is less clear.[10] The hegemony of computationalism is not total though, and there have been a number of opposing voices in the decades since cognitive science began, not least of which comes from the various strands of enactivism.

9.1.2 Enactivism

As we shall see in the next section, the anti-computationalist tradition that has become known as enactivism has its origins in (second-order) cybernetics. Its foundational text is Varela, Thompson and Rosch's (1991) *The Embodied Mind*, where the term 'enactivism' is first used to refer to a theory of mind that rejects a representational characterisation of cognition in favour of the idea that meaning arises (or is 'enacted') out of the interaction between mind and world. Varela, Thompson and Rosch's original presentation of enactivism drew on various sources of inspiration, including phenomenology, Buddhist philosophy, and Maturana's 'autopoietic' theory of cognition, which characterises cognition as a process of autonomous 'self-production'. This chapter will focus primarily on the latter influence, which I will trace back to the cybernetic notion of homeostasis, before making the relatively novel argument that enactivism can actually be made compatible with a computational characterisation of cognition.[11] This argument is based on an examination of the historical roots of both approaches to cognitive science, demonstrating the importance of an iHPS approach.

Ward, Silverman and Villalobos (2017) distinguish between three main varieties of contemporary enactivism: 'autopoietic', 'sensorimotor', and 'radical'. Autopoietic enactivism emphasises the 'autonomous' nature of biological systems, drawing on the autopoietic theory of Maturana (and Varela). The approach originated in Varela, Thompson, and Rosch's (1991) *The Embodied Mind*, and contemporary proponents include Ezequiel Di Paolo (de Jaegher and di Paolo 2007), Xabier Barandiaran (2006), and Evan Thompson (2007) himself. Sensorimotor enactivism emphasises the role of action in perception and cognition, via the notion of a 'sensorimotor contingency', in other words, an awareness of what would happen (perceptually) if we were to move or act in a certain way. Contemporary proponents of this approach include Kevin O'Regan and Alva Noë (2001), and Susan Hurley (2002). Finally, radical enactivism, recently proposed by Daniel Hutto and Eric Myin (2012), emphasises the rejection of any notion of intrinsic representational content. All three varieties of enactivism share a general anti-cognitivist attitude, while at the same time disagreeing on the exact details of their positive proposals. In particular, sensorimotor enactivism need not necessarily reject computational or representational characterisations of cognition, whereas both autopoietic and radical enactivism share an opposition to mental representation, which contributes to their rejection of computationalism.

For autopoietic enactivism, the rejection of mental representation necessarily entails a rejection of computationalism. The primary reason for this is that the autopoietic enactivists have historically assumed that computation requires representation, in line with the classical computational theory of mind that they were responding to: if computation requires representation, and enactivism rejects representation, then it is quite clear that enactivism must also reject computation. However, the opposition goes slightly deeper than this, and can also be found in their initial reasons for rejecting the idea of mental representation. These have to do with the emphasis that they place on the autonomy of living (cognitive) systems, which the autopoietic enactivists think rules out the characterisation of cognitive systems as representational systems. This autonomy requirement also speaks directly to their rejection of computationalism, as the various criteria for autonomy that they endorse are thought to be incompatible with computation.[12] For example, an autonomous system is said to be 'self-determined', ruling out the kinds of instructional inputs that a computer is typically understood to require. Thompson (2007, p. 37) explicitly contrasts autonomous systems of this kind with 'heteronomous systems', such as an 'automatic bank machine', whose activities are 'determined and controlled from the outside'. So computation is ruled out not only because it is thought to require representation, but also because a computational system is thought to be a paradigmatic example of a heteronomous system, placing it in direct opposition to the autonomy requirement. I will return to the question of autonomy and computation towards the end of the chapter, where I will suggest that this opposition is not quite as clear as it first appears. First, however, I will trace the cybernetic origins of both autonomy and computation, in order to give some initial motivation for the idea that they might in fact be compatible.

9.2 Cybernetic origins

Both enactivism and computationalism have their origins in the cybernetics movement of the mid-twentieth century. This movement, which was inspired by technological developments during World War II, had its heyday around a series of interdisciplinary conferences (the 'Macy conferences') held in the USA from 1946 to 1953.[13] The movement began to splinter after the final conference in 1953, with many researchers going on to become influential in what would become mainstream (computational) cognitive science, while others formed the 'second-order' cybernetics of the 1960s and 1970s. This latter group would eventually influence and inspire the first enactivists, who drew on second-order cybernetics to present their ideas. In this section I will consider two themes from the original cybernetics movement: neural computation and biological homeostasis. I will argue that each of these contain the seeds of what would eventually become computationalism and enactivism (respectively), and thus exemplify an era when the two now-hostile approaches to cognitive science were able to cooperate.

The term 'cybernetics' was coined by the mathematician Norbert Wiener as a neologism from the Greek *kybernetes*, or 'steersman', evoking the idea of a system being purposefully controlled. He subsequently defined cybernetics as 'the scientific study of control and communication in the animal and the machine'.[14] This new science was inspired by his experiences designing anti-aircraft gun targeting systems during World War II, and was intended to bring together various strands of mathematics, physics, biology, electrical engineering, and other disciplines. Wiener thought that the new science of cybernetics would provide an explanation of the complex behaviour of certain (seemingly intelligent) systems, both natural and artificial. In an earlier article, written in collaboration with neurophysiologist Arturo Rosenblueth and engineer Julian Bigelow, Wiener had argued that goal-directed behaviours, such as that of a guided torpedo correcting its course or an animal seeking out food, could be explained in terms of a general theory of negative feedback.[15] They thus sought to unify the study of both natural and artificial 'teleological' systems, at the same time as reincorporating them into a naturalistic or deterministic worldview. This general insight or approach would provide the foundation for the cybernetics movement as a whole.

The Macy conferences, initially organised by Lawrence K. Frank and Frank Fremont Smith, and sponsored by the Josiah Macy, Jr. Foundation, brought together a diverse range of researchers with the general aim of developing a science of the human mind.[16] Ten conferences were held between 1946 and 1953, and the ideas cultivated in them would provide the foundations for what would eventually become cognitive science. The conferences each had an official title, usually referring to 'circular feedback mechanisms', but it was not until the seventh conference that the term 'cybernetics' was included in the title.[17] The talks and ensuing discussions at the conferences covered a vast range of topics, which all shared, as a common point of reference, the aim of trying to understand the human mind in ways that were amenable to contemporary scientific practice.

In the rest of this section I will consider two key themes that arose during the conferences, neural computation and biological homeostasis, which correspond approximately to two themes that Jean-Pierre Dupuy (2009, pp. 3–4) identifies in early cybernetics, 'thinking is a form of computation' and 'physical laws can explain the appearance of meaning in the world'. I will focus on these themes because each went on to inspire, respectively, the computational and enactive theories of mind, which are now seen as being diametrically opposed to one another. The connection between the cybernetic notion of neural computation and the contemporary computational theory of mind is quite straightforward, and I will not say much more about it here.[18] Roughly speaking, if you think that the brain is a computer, and that mental activities or processes might also be computational, then it becomes easy to see what the relationship between mind and brain might be (something like the relationship between software and hardware in artificial computational systems).

9.2.1 Neural computation

The idea of treating neural activity, in the form of synaptic firing, as instantiating a form of propositional logic, was first proposed by Warren McCulloch and Walter Pitts (1943) in their article 'A Logical Calculus of the Ideas Immanent in Nervous Activity'. In this article they describe how the 'all-or-none' character of neuronal activity means that the neuron can be treated as essentially equivalent to a computational component performing a logical function. A neuron receives electrical impulses from preceding neurons, which serve to 'excite' the neuron until it reaches a certain threshold, at which point it releases chemicals ('neurotransmitters') into the synaptic gap, generating an electrical impulse in the subsequent neuron, and continuing the process.[19] A neuron with two preceding ('input') neurons, a threshold that requires a simultaneous (or at least temporally proximal) impulse from both, and one subsequent ('output') neuron, could therefore be treated as performing the logical operation AND, where an electrical impulse is treated as a '1', and the lack of any such impulse as a '0'. A 'net' of such neurons can be treated as performing more complex operations, allowing for a general characterisation of neural activity as a 'logical calculus'.

Towards the end of their paper, McCulloch and Pitts (1943, p. 17) claim that a neural net of this kind is equivalent in computational power to a Turing machine. A Turing machine, as described by the mathematician and early computer scientist Alan Turing (1936), is a simple device consisting of an automaton with sensors, effectors, and a motor, supplied with a tape that it can mark with a number of discrete symbols (canonically, '0' and '1'). The automaton reads a symbol off the tape, and then, according to rules defined by its internal structure, systematically either replaces the symbol or moves to the next symbol on the tape, and so on. Turing's great insight was that even this very simple system is capable of performing a vast number of computations, in fact providing a definition of 'computability' that is equivalent to several other proposed definitions. According to McCulloch and Pitts' characterisation of neural activity, the brain is essentially a Turing machine, and thus a kind of general purpose computer. Their paper therefore marks the birth of the computational theory of mind, as it provides a way of describing neural activity, and the behaviour it produces, in terms of computational processing.[20]

John von Neumann, an occasional attendee of the Macy conferences and the original designer of the basic architecture common to all modern computers, took inspiration from McCulloch and Pitts in his own work on neural computation.[21] In *The Computer and the Brain*, von Neumann (2000) explores the possibility of a computational theory of the mind, comparing the processing power of an artificial electronic computer with that of the brain. Finding the latter to be somewhat inferior, he suggests that some other mechanisms or properties of the brain must be relevant to cognition, such as the fine-grained structure of the neuron itself or the diffusion of chemical signals. He also argues that the brain might be using a distinctive 'statistical language', rather than the digital system assumed by

McCulloch and Pitts.[22] Despite his disagreement with parts of their paper, he was still content to describe cognition as being produced in some sense by the computational properties of the brain, whatever these may be.

The cybernetic understanding of neural computation, as expressed by McCulloch and Pitts and von Neumann, makes no reference to representational content, in contrast with the computational theory of mind that subsequently developed. Insofar as it invoked anything like information, it was the meaningless, non-representational information described by Claude Shannon's mathematical theory of information.[23] Thus, as Dupuy (2009, p. 7) notes, 'computation was first introduced into the construction of a materialist and physicalist science not as symbolic computation involving representations, but instead as a sort of blind computation having no meaning whatever, either with respect to its objects or to its aims'. This feature, as we shall see in the final section, could make the cybernetic notion of neural computation somewhat more amenable to the enactivists.

9.2.2 Biological homeostasis

The cybernetic concept of homeostasis has its origins in the work of Walter Cannon (1929), who coined the term in 1926 to describe the biological mechanisms responsible for maintaining the stability or integrity of a living organism. There are many mechanisms of this kind, including, for example, the maintenance of blood-glucose levels by consuming food and releasing insulin, and the maintenance of blood-oxygen levels via respiration. These mechanisms also interact in interesting and complex ways, with the overall result of keeping the organism alive and healthy for as long as possible.

Wiener (1948, p. 114) briefly mentions homeostasis, relating it to the more general notion of 'feedback', but does not discuss it in any detail. The cyberneticist most associated with homeostasis is W. Ross Ashby (1960), who draws on the concept to help illustrate his general definition of an 'adaptive' system as one which naturally returns to a stable (or homeostatic) configuration. He has in mind here a somewhat broader class of mechanisms than Cannon, including not only basic biological systems but also the operation of various artefacts created by humans and other animals. Ashby (1960, p. 62) was interested in the potential for this kind of explanation to be 'applied to a great deal, if not all, of the normal human adult's behaviour', including learning and cognition more generally. Furthermore, it seemed to offer Ashby a way of naturalising the concept of 'purpose' or 'meaning': if the purpose of a system is simply to achieve homeostasis, then it could legitimately be described as 'aiming' for this goal, without invoking any external source of meaning.

Something like this approach can also be found in Rosenblueth, Wiener and Bigelow's (1943) earlier work, 'Behavior, Purpose, and Teleology'. They do not refer explicitly to homeostasis, but one of the authors (Rosenblueth) worked in Cannon's laboratory at Harvard, and their definition of purpose in terms of circular feedback certainly calls to mind Ashby's later discussion in

Design for a Brain.[24] They describe how any 'self-correcting' system, whether natural or artificial, can be understood in terms of negative feedback, which is feedback that adjusts the current trajectory of the system, rather than simply encouraging it to continue as it is. The behaviour of a system of this kind can appear purposeful to an observer, as knocking it 'off-course' will generate feedback that puts it back 'on-course', thus making it behave as though it has a definite goal in mind. For Rosenblueth, Wiener and Bigelow (1943, pp. 19–22), this is more than an appearance, and we should simply take behaviour of this kind to be genuinely purposeful, albeit at various levels of complexity and significance. Thus, like Ashby, they sought to explain the complex behaviour of an organism (or mechanism) in terms of the simpler notion of feedback and/or homeostasis.

Both of the themes explored in this section (neural computation and biological homeostasis) coexisted for a time in the early cybernetics. A nice illustration of this coexistence are the autonomous robots produced by W. Grey Walter, which utilised computational circuitry in order to maintain a form of homeostatic equilibrium.[25] These notions of homeostasis and computation later came to be seen as standing in opposition to one another, or at least not obviously related. Computational theories of mind became increasingly abstract and distanced from the physical realities of biological cognition, while homeostasis is a fundamentally biological notion, which only makes sense in the context of an embodied and environmentally situated organism.[26] Dupuy (2009, p. 11) clearly expresses this opposition, writing that:

> the first cybernetics, confronted with the theories of self-organization and complexity that were to be dear to its successor [second-order cybernetics and eventually enactivism], turned its back on them, and indeed sometimes – cruel irony! – actually combatted them.

Mainstream, computational cognitive science became hostile to the approaches developed by second-order cybernetics, and contemporary histories of cognitive science typically have little to say about the original cybernetics movement.[27] In the next section I will describe how the cybernetic notion of homeostasis was instrumental to the development of the enactivist notion of autonomy, via Maturana's idea of 'autopoiesis'. Understanding this development is crucial if we want to see how and why enactivism came to reject computationalism, and how we might be able to reconsider this rejection in light of a more modern understanding of computation (albeit one that echoes the original cybernetics).

9.3 The enactivist notion of autonomy

There is much that could be said about the relationship between contemporary enactivism and classical cybernetics.[28] I want to focus here on just one aspect of that relationship, the development of the enactivist notion of autonomy out of the cybernetic notion of homeostasis, as it is this aspect that casts most light on the contemporary dispute between enactivism and computationalism. Autonomy, as we shall see in the next section, is central to the enactivist rejection of computationalism, and so in understanding its origins we can begin to explore the possibility of a reconciliation between the two approaches.

9.3.1 Homeostasis

For the original cyberneticists, the homeostatic system provided a general model of both life and cognition. A living organism is a homeostatic system, or perhaps a collection of several homeostatic systems which jointly contribute to the continued existence of the organism. Such systems will eventually collapse, and thus cease 'living', but so long as they are able to maintain homeostatic stability we can consider them to be alive. Add an additional layer of complexity, in the form of what Ashby calls 'ultrastability', or the capacity to modify one's own internal or external environment in ways that are conducive to homeostasis, and we get something approaching intelligence or cognition.[29] Ashby illustrated his model of homeostasis with a literal 'homeostat', a simple network of four electrical devices, each connected to one another, but each also seeking to maintain its own homeostasis.[30] The devices could be set to a random configuration, and after a period of (apparently) chaotic fluctuations would eventually settle into a stable configuration. Ashby's presentation of his homeostat at the ninth Macy conference was somewhat controversial, and perhaps helped precipitate the collapse of the first cybernetics movement, but this simple idea went on to inspire what would become enactivism.[31]

9.3.2 Autopoiesis

As I mentioned earlier, the link between the original cybernetics movement and the modern enactivist tradition can be found in the works of the Chilean biologist and philosopher Humberto Maturana. Although Maturana was only loosely connected to the original cybernetics movement through his collaborative work with McCulloch and Pitts,[32] he took inspiration from Ashby's work in the presentation of his autopoietic theory of cognition. This theory proposed that the defining feature of cognition (and life) is what he called 'autopoiesis', another Greek neologism intended to mean something akin to 'self-production'.[33] In the *Biology of Cognition*, reprinted as part one of *Autopoiesis and Cognition*, Maturana (1970) makes frequent reference to homeostasis as essential to the definition of the living organism. Although at this point he had not yet coined the term 'autopoiesis', the connection between the two concepts should be apparent. Maturana's archetypal example of an autopoietic system was a biological cell, which contains all of the information and mechanisms necessary to sustain its own existence, at least until it suffers irrevocable damage or otherwise collapses. Likewise, a homeostatic system is able to maintain its own structural integrity in response to external disturbances. Living organisms, according to Maturana's theory, are composed of a hierarchy of autopoietic structures, nested within one another. Cognition, at least in Maturana's earlier work, is seen as being continuous with life, and what we call a cognitive system is simply a living system of sufficient complexity.[34]

9.3.3 Autonomy

Maturana's student and co-author Francisco Varela in turn developed the notion of autopoiesis into the enactivist notion of autonomy. The enactivist notion of autonomy consists of three aspects: self-determination, organisational/ operational closure, and thermodynamic precariousness. As mentioned briefly in the first section, these three criteria are intended to be (partially) constitutive of life and cognition, and for reasons that I will address in the next section, have historically been assumed to rule out computational theories of mind. In the remainder of this section I will describe each aspect of autonomy, and how they relate to cybernetics and autopoietic theory, in order to elucidate the way in which enactivism developed out of ideas originating in the cybernetics movement.

The first element of the enactivist notion of autonomy is self-determination, which refers to the way in which the behaviour of an autonomous system is determined by its own structural dynamics, rather than by external instruction or control. We might think here of the distinction between a vehicle operated by a driver, behaving only according to their instructions, and a living organism, which seems to behave according to its own self-generated dynamics. However, for Maturana (1987, p. 73) at least, something like self-determination is in fact true of all physical systems, insofar as their responses to external stimuli are determined according to their own physical structure. This becomes less clear in later enactivist writings, which suggest that self-determination will only apply to certain kinds of systems, and could thus serve as part of a definition of life or cognition.

The second element has two aspects: organisational closure refers to what Thompson (2007, p. 45) describes as 'the self-referential (circular and recursive) network of relations that defines the system as a unity' and operational closure to 'the re-entrant and recurrent dynamics of such a system'. The point here is that the causal dynamics which are significant to an autonomous system should be seen to return to that system through its environment, forming a closed loop. Each unit of Ashby's homeostat exhibits organisational/operational closure insofar as the effect it has on the other units returns to it via the recursive effect that they in turn have on the original unit. In terms of a living organism, we can think of the way that any significant action it takes will be one that modifies its environment so as to change the effect that that environment has on the organism itself. Another way of putting this is that the sensorimotor dynamics of a living organism form a closed loop with its environment, with every action provoking a reaction that produces a novel sensory experience. This aspect of autonomy has its roots in Maturana's notion of functional closure, which again can be understood to apply more broadly than just to living organisms.[35]

The final element is thermodynamic precariousness, which refers to the status of living organisms as dissipative structures, that is to say, a special group of systems that maintain their structural organisation by consuming energy from their environment, rather than by simply resting at a point of stable thermodynamic equilibrium.[36] This echoes Ashby's distinction between stability and ultrastability, and Maturana's characterisation of an autopoietic (or living) system as one that is able to make use of energy from its environment in order to maintain its own structure. According to the enactivist definition, an autonomous system must also be thermodynamically precarious, such that it is able to (temporarily) counteract the inevitable tendency towards thermodynamic decay.[37]

These three aspects of the more general autonomy requirement form only one part of the enactivist story, but as we will see in the next section it is this requirement that seems to speak most clearly against the computational characterisation of cognition. I will first demonstrate why these criteria have been thought to rule out computationalism, before arguing that there is in fact a way of understanding computation which can be made compatible with autonomy.

9.4 Autonomy and computation

In the previous section we saw how the enactivist notion of autonomy developed out of the cybernetic notion of homeostasis. In this final section I will first outline the semantic account of computation, which underpins the classical computational theory of mind, before demonstrating why this account of computation is incompatible with the enactivist notion of autonomy. Finally, I will introduce a more recent mechanistic account of computation, which I will suggest could be made compatible with enactivism, and thus could offer one route towards a future reconciliation between enactivism and computationalism.

9.4.1 The semantic account of computation

The semantic account of computation developed out of work by Putnam and Fodor in the early days of cognitive science (see section 9.1). In the context of Fodor's project of 'naturalising' folk psychological categories like belief and desire, it made sense to think of the computational structure that was meant to implement these categories as representing features of the world. The computational state implementing my belief that the sky is blue is naturally interpreted as representing how the world actually is, while the computational state implementing my desire for more coffee is naturally interpreted as representing how I want the world to be. Therefore, there seemed to be a natural fit between Fodor's representational theory of mind and the computational theory of neural processing that Putnam inherited from the cyberneticists.

These attributions of representational content are not just natural interpretations, however: according to the semantic account they serve a further role in individuating computational states and processes. This only really became apparent after a major challenge to the computational theory of mind posed (in slightly different forms) by the philosopher John Searle and (ironically) Putnam himself. Both Searle and Putnam demonstrated that, when understood purely syntactically or structurally, almost any physical system could be interpreted as performing a wide range of computations, leading to the potentially absurd conclusion that everything is a computer.[38] In brief, the problem is that it is quite simple to describe a relationship between any arbitrary configuration of physical states and any (or at least some) computational operation, such that those physical states can be interpreted as performing that operation. If *everything* computes in this sense, then it is no longer especially interesting to say that *any particular thing* computes, thus rendering the computational theory of mind somewhat trivial.[39]

There are many ways of responding to this challenge, more than I have space to describe here, but one particularly popular response is to add an additional 'semantic' criterion to computation.[40] According to this criterion, instantiating a computation requires not only the correct physical structure, but also that this structure represents the world in the correct manner. Again, the details of this account can be spelled out in several different ways, but the key point is that it is committed to the idea that physical computations should be considered intrinsically 'meaningful', because they in some sense represent features of the external world. It is precisely this idea of intrinsic representational content that the enactivist tradition rejects, as they think that it threatens the autonomy of the cognitive system. To say that a system represents its environment implies (for the enactivist) a 'correct' way in which this environment ought to be represented, requiring an intentional agency outside of the system in order to establish these representational norms. In the case of artificial computers, designed and created by human agency, this need not pose any problem, as these systems can be seen as representing their environments in virtue of how we have chosen to build them (their representational status is in this sense 'parasitic' on our own intentional capacities). However, the representational (or semantic) account of computation becomes more problematic when applied to biological systems, which do not typically have a 'designer' in the same sense that an artificial computer does.[41] Enactivists have therefore historically assumed that cognition cannot be computational, on the basis that computation must be representational and cognition (according to the autonomy requirement) cannot be. In the next section I will describe the enactivist rejection of computation in slightly more detail, before finally suggesting that contemporary 'mechanistic' accounts of computation, which are not necessarily representational, could in principle be made compatible with enactivism and the autonomy requirement.

9.4.2 The enactivist rejection of computation

In order to understand the enactivist rejection of computationalism, it is important to recognise that enactivism has taken for granted a semantic or representational account of computation, as described above. According to Varela, Thompson and Rosch's original presentation of enactivism, computations are thought to be intrinsically representational, and thus incompatible with the autonomy requirement.[42] Due to their representational status, computational systems are seen as being dependent on an external, 'pre-given' world, while autonomous systems do not represent an external world, and are instead able to '*enact* a world as a domain of distinctions that is inseparable from the structure embodied by the cognitive system'.[43] The key point here is that for an autonomous system, the world is only meaningful insofar as that meaning is generated by the system's interaction with its environment. In contrast, a heteronomous system, such as a semantically defined computer, is only able to pick up on meaning that already exists in the world. As the latter is incompatible with the enactivist theory of mind, they reject the semantic characterisation of cognition, and thus also the computational characterisation of cognition (insofar as computation is understood to be a semantic phenomenon).

Each aspect of the autonomy requirement also seems to speak against the possibility of a computational theory of mind. Computers, at least as they are classically understood, respond to semantically or informationally rich instructions, and are therefore not self-determined. They also receive discrete inputs and outputs, which may bear no direct relation to one another, and are therefore not organisationally or operationally closed. Finally, they typically rely on external maintenance in order to retain structural stability, and are therefore not thermodynamically precarious or homeostatically ultrastable. All three aspects of the autonomy requirement seem, for the enactivists, to speak against computational characterisations of living or cognitive systems.[44]

9.4.3 The mechanistic account of computation

While it seems clear that computation, according to the semantic account, is incompatible with the autonomy requirement and the more general enactivist theory of mind, this incompatibility does not seem so obvious for more recent 'mechanistic' accounts of computation. These mechanistic accounts have been defended most comprehensively by the philosopher Gualtiero Piccinini, whose work I focus on here.[45]

According to Piccinini's mechanistic account of computation, a physical computer is a mechanism whose function is to perform systematic transformations over medium independent variables.[46] Note that this definition makes no reference to any form of semantic or representational content, and yet it is nonetheless able to distinguish computational from non-computational systems, thus avoiding triviality or pancomputationalism. The account therefore seems *prima facie* compatible with the enactivist autonomy requirement, as it avoids the semantic characterisation that enactivism was primarily opposed to. It also seems compatible with each aspect of autonomy, although some minor modifications to Piccinini's account may be required. A computing mechanism is self-determined insofar as its responses to external disturbances are determined solely by its own physical structure, with no reference to external instruction or control.[47] It can be organisationally/operationally closed if it is arranged in such a way that the effect it has on the world loops back through the environment to impact its sensory surfaces.[48] Finally, it can be thermodynamically precarious if it is designed such that it can seek out a suitable source of energy and harness this energy in order to maintain its own structural integrity.[49] It is important to note that the latter two aspects will only be true of a certain special class of computing mechanism. The claim here is not that *all* computing mechanisms are autonomous systems, but rather that *some* computing mechanisms could be, and therefore that there is no in-principle reason why cognition could not be characterised as both computational (in the mechanistic sense) and also autonomous/enactive. A computational characterisation of neural processing is therefore not necessarily incompatible with an enactivist theory of mind, potentially healing the rift that entered cognitive science after the collapse of the original cybernetics movement.

Conclusion

By tracing the shared origins of enactivism and computationalism in the cybernetics movement of the 1940s and 1950s, I hope to have demonstrated that these two approaches to cognitive science need not be seen as inevitably opposed to one another, and could even fruitfully combine their efforts in future cognitive scientific research. There is much more work that needs to be done before this combined approach becomes a reality, but already we are seeing the resurgence of research in cognitive science that takes inspiration from both traditions. The emerging 'free energy minimisation' framework provides an example of a cognitive scientific theory that appeals explicitly to the original cybernetics, combining homeostatic mechanisms with a form of (possibly non-semantic) computational processing.[50] Recent work by Matteo Colombo and Cory Wright takes a historical approach to this framework, as does the introductory chapter to an edited volume on its philosophical implications.[51] I am hopeful that this trend will continue in the future, and that an integrated approach to the History and Philosophy of Science will be able to provide insights into contemporary cognitive scientific issues and debates.[52]

Notes

1 Examples of anti-computationalist enactivists include Francisco Varela et al., (1991); Evan Thompson (2007); and Daniel Hutto and Eric Myin (2012).
2 Computationalism in philosophy of mind is the view that the mind (or brain) is a computer, and that mental states and processes are computational states and processes. Enactivism refers to a family of views that have historically rejected the computational characterisation of cognition in favour of the idea that mental

activity arises out of the interaction between brain, body, and world. Both approaches are described in more detail in the next section.

3 Autopoietic theory emphasises the biological nature of cognition, and was a precursor to enactivism. It is described in more detail in section 9.3.2.

4 For a philosophical overview of computational approaches to cognitive science, see *The Routledge Handbook of the Computational Mind*, ed. Mark Sprevak and Matteo Colombo (2018). For a recent defence of the progressiveness of this research programme, see Marcin Milkowski (2018).

5 Milkowski (2018).

6 Michael Rescorla, n.d.

7 Dave Ward, et al. (2017) (section 9.3).

8 Hilary Putnam (1967). See also Rescorla, n.d. section 3.1.

9 For an introduction to computational neuroscience, see Thomas P. Trappenberg (2009). For an overview of recent debates about neurocognitive ontologies, see Michael Anderson (2015).

10 See Sprevak and Colombo (2018) for extensive further discussion of these issues.

11 I have previously developed versions of this argument in collaboration with Mario Villalobos, most comprehensively in Mario Villalobos and Joe Dewhurst (2017a).

12 Villalobos and Dewhurst (2017a) , section 2.

13 There was a parallel development of a distinctive 'British' cybernetics that I will not discuss in any detail here, aside from mentioning Ashby's work on homeostasis. For a more detailed discussion of British cybernetics, see Andrew Pickering (2010); for an analysis of the relationship between the British cyberneticists and computational theories of mind, see Joe Dewhurst (2018).

14 Norbert Wiener (1948).

15 Arturo Rosenblueth, Norbert Wiener, and Julian Bigelow (1943).

16 For an overview of the conferences, including a full list of speakers and attendees, see Steve J. Heims (1993).

17 The conferences were officially titled as follows: First conference; 'Feedback Mechanisms and Circular Causal Systems in Biological and Social Systems', second and third conferences: 'Teleological Mechanisms and Circular Causal Systems', fourth, fifth, and sixth conferences: 'Circular Causal and Feedback Mechanisms in Biological and Social Systems', seventh and all subsequent conferences; 'Cybernetics: Circular Causal and Feedback Mechanisms in Biological and Social Systems'.

18 For further discussion see Tara Abraham (2018).

19 This is an oversimplification, but should suffice to illustrate the point. For an introductory discussion of neural coding, see Rosa Cao (2018).

20 This idea also has an important 'pre-history', described by Alistair Isaac (2018). For further discussion see Gualtiero Piccinini (2004).

21 The so-called 'von Neumann architecture' is first described in John von Neumann (1945). For further discussion see William Aspray (1990).

22 This suggestion anticipates the development of connectionism and artificial neural networks, some three decades later.

23 Whilst McCulloch and Pitts developed their account of neural computation independently of Shannon's theory of information, the latter certainly had an influence on the cybernetics movement as a whole. For a discussion of the complex relationship between computation and information in cybernetics and beyond, see Gualtiero Piccinini and Andrea Scarantino (2010). For Shannon's original theory of information, see Claude Shannon (1948). For an argument that Shannon's information is in fact semantic, see Alistair Isaac (2019).

24 Dupuy (2009), p. 45.

25 W. Grey Walter (1950). Walter's robots are discussed in more detail by Dewhurst (2018, sec. 3), and by Owen Holland (2003).

26 This is not to say that there can be no such thing as an artificial homeostat, such as Ashby's own creations, or a virtual simulation of homeostasis, but rather that all such systems are comparable to a biological organism in ways that an abstract computer program is not.

27 For a notable exception, see Margaret Boden (2006), especially chapter 4.

28 See for example Tom Froese (2010, 2011).

29 Ashby (1960), chapter 7.

30 W. Ross Ashby (1948); cf. Boden (2006), p. 230.

31 Boden (2006), pp. 232–235; Dupuy (2009), p.150.

32 Jerome Lettvin et al. (1959). Maturana was also associated with the 'second-order cybernetics' discussed by Froese (2010, 2011).

33 Humberto Maturana and Francisco Varela (1980), p. 16.

34 Maturana and Varela (1980), p. 13.

35 Humberto Maturana (1975); Mario Villalobos (2015).

36 Mario Villalobos (2016).

37 Ezequiel di Paolo and Evan Thompson (2014), p. 72,

38 John Searle (1992); Hilary Putnam (1988). For an overview of these arguments, see Gualtiero Piccinini (2015), chapter 4.

39 It is worth noting that some philosophers have simply accepted this apparently absurd conclusion, while maintaining that a computational description of this sort might still be interesting or useful. See for example Paul Schweizer (2014).

40 For an overview of semantic accounts of computation, see Piccinini (2015), chapter 3.

41 It is not unusual to hear people speaking of evolution or natural selection as the 'designer' of biological systems, but in most cases this way of speaking is clearly metaphorical, or at least distinctly non-intentional.

42 Varela, Thompson, and Rosch (1991), pp.140–141.

43 Varela, Thompson, and Rosch (1991), pp.135–140.

44 Villalobos and Dewhurst (2017a), section 2.

45 The arguments in this section summarise recent work that is developed in more detail in Villalobos and Dewhurst (2017a, 2017b); and Joe Dewhurst and Mario Villalobos, (2017).

46 Piccinini (2015), chapter 7.

47 Villalobos and Dewhurst (2017a), section 6.

48 Villalobos and Dewhurst (2017a), section 5.

49 Villalobos and Dewhurst (2017a), section 7.

50 Karl Friston (2013); Anil Seth (2015). For an analysis of the representational status of this framework, see Krzysztof Dolega (2017).

51 Matteo Colombo and Cory Wright (2018); Wanja Wiese and Thomas Metzinger (2017).

52 Dupuy, Boden, and Froese each develop such an approach, as do several of the chapters in the first section of Sprevak and Colombo (2018).

Acknowledgements

An earlier version of this chapter was presented at the *Sixth International Conference on Integrated History and Philosophy of Science* at the University of Edinburgh in July 2016, and I am grateful to the organisers and audience members at that conference for their feedback. I am also grateful to Mario Villalobos and Dave Ward, whose collaboration and support has been instrumental in shaping my understanding of enactivism and autopoietic theory. Finally, I would like to thank the organisers of *The Past,*

Present and Future of Integrated HPS for hosting such an inspiring conference, and the reviewers for their helpful and encouraging comments.

Bibliography

Abraham, Tara, 'Cybernetics', in *The Routledge Handbook of the Computational Mind*, Mark Sprevak and Matteo Colombo, eds. (London: Routledge, 2018).

Anderson, Michael, 'Mining the Brain for a New Taxonomy of the Mind', *Philosophy Compass*, 10. 1(2015), 68–77.

Ashby, W. Ross, 'Design for a Brain', *Electronic Engineering*, 20(1948), 379–383.

Ashby, W., *Design for a Brain*, 2nd edn (New York: John Wiley & Sons, 1960).

Aspray, William, *John von Neumann and the Origins of Modern Computing* (Cambridge, MA: MIT Press, 1990).

Barandiaran, Xabier, 'On What Makes Certain Dynamical Systems Cognitive', *Adaptive Behavior*, 14. 2(2006), 171–185.

Boden, Margaret, *Mind as Machine* (Oxford: Oxford University Press, 2006).

Cannon, Walter, 'Organization for Physiological Homeostasis', *Physiological Reviews*, 9. 3(1929), 399–431.

Cao, Rosa, 'Neural Coding', in *The Routledge Handbook of the Computational Mind*, ed. by Mark Sprevak and Matteo Colombo (London: Routledge, 2018).

Colombo, Matteo, and Cory Wright, 'First Principles in the Life Sciences: The Free-energy Principle, Organicism, and Mechanism', *Synthese*, online first (2018), https://doi.org/10.1007/s11229-018-01932-w/

De Jaegher, Hanne, and Ezequiel di Paolo, 'Participatory sense-making', *Phenomenology and the Cognitive Sciences*, 6. 4(2007), 485–507.

Dewhurst, Joe, 'British Cybernetics', in *The Routledge Handbook of the Computational Mind*, ed. by Mark Sprevak and Matteo Colombo (London: Routledge, 2018).

Dewhurst, Joe and Villalobos, Mario, 'The Enactive Automaton as a Computing Mechanism', *Thought*, 6. 3(2017), 185–192.

di Paolo, Ezequiel and Thompson, Evan, 'The Enactive Approach', in *The Routledge Handbook of Embodied Cognition*, ed. by Larry Shapiro (New York: Routledge Press, 2014).

Dolega, Krzysztof, 'Moderate Predictive Processing', in *Philosophy and Predictive Processing* (2017), available at https://predictive-mind.net/papers

Dupuy, Jean-Pierre, *On the Origins of Cognitive Science*, translated by M. B. DeBevoise, 2nd edition (Cambridge, MA: MIT Press, 2009).

Fodor, Jerry, *The Language of Thought* (Cambridge, MA: Harvard University Press, 1975).

Friston, Karl, 'Life as We Know It', *Journal of the Royal Society Interface*, 10(2013).

Froese, Tom, 'From Cybernetics to Second-Order Cybernetics', *Constructivist Foundations*, 5. 2(2010), 75–85.

Froese, Tom, 'From Second-Order Cybernetics to Enactive Cognitive Science', *Systems Research and Behavioural Science*, 28. 6(2011), 631–645.

Heims, Steve J., *The Cybernetics Group 1946–1953* (Cambridge, MA: MIT Press, 1993).

Holland, Owen, 'The First Biologically Inspired Robots', *Robotica*, 21(2003), 351–363.

Hurley, Susan, *Consciousness in Action* (Cambridge, MA: Harvard University Press, 2002).

Hutto, Daniel, and Eric Myin, *Radicalizing Enactivism* (Cambridge, MA: MIT Press, 2012).

Isaac, Alistair, 'The Semantics Latent in Shannon Information', *British Journal for the Philosophy of Science*, 70, 1(2019), 77–102.

Isaac, Alistair, 'Computational Thought from Descartes to Lovelace', in *The Routledge Handbook of the Computational Mind*, ed. by Mark Sprevak and Matteo Colombo (London: Routledge, 2018).

Lettvin, Jerome, Humberto Maturana, Warren McCulloch, and Walter Pitts, 'What the Frog's Eye Tells the Frog's Brain', *Proceedings of the IRE* (1959), 1940–1951.

Maturana, Humberto, *Biology of Cognition* (Urbana, IL: University of Illinois, 1970).

Maturana, Humberto, 'The Organization of the Living', *International Journal of the Man-Machine Studies*, 7(1975), 313–322.

Maturana, Humberto, 'Everything is Said by an Observer', in *GAIA: A Way of Knowing*, ed. by W. I. Thompson, (Hudson, NY: Lindisfarne Press, 1987).

Maturana, Humberto and Francisco Varela, *Autopoiesis and Cognition: the Realisation of the Living*, 2nd edn (Dordecht: D. Reidel Publishing Company, 1980).

McCulloch, Warren and Walter Pitts, 'A Logical Calculus of the Ideas Immanent in Nervous Activity', *Bulletin of Mathematical Biophysics*, 5(1943), 115–133.

Milkowski, Marcin, 'From Computer Metaphor to Computational Modeling: The Evolution of Computationalism', *Minds & Machines*, 28. 3(2018), 515–541. https://doi.org/10.1007/s11023-018-9468-3

O'Regan, Kevin, and Alva Noë, 'A Sensorimotor Account of Vision and Visual Consciousness', *Behavioral and Brain Sciences*, 24(2001), 939–1031.

Piccinini, Gualtiero, 'The First Computational Theory of Mind and Brain', *Synthese*, 141(2004), 175–215.

Piccinini, Gualtiero, *Physical Computation: A Mechanistic Account* (Oxford: Oxford University Press, 2015).

Piccinini, Gualtiero and Andrea Scarantino, 'Computation vs. Information Processing: Why Their Difference Matters to Cognitive Science', *Studies in the History and Philosophy of Science Part A*, 41. 3(2010), 237–246.

Pickering, Andrew, *The Cybernetic Brain* (Chicago: University of Chicago Press, 2010).

Putnam, Hilary, 'Psychological Predicates', in *Art, Philosophy, and Religion* (Pittsburgh, PA: University of Pittsburgh Press, 1967).

Putnam, Hilary, *Representation and Reality* (Cambridge, MA: MIT Press, 1988).

Rescorla, Michael, 'The Computational Theory of Mind', in *The Stanford Encyclopedia of Philosophy*, ed. by Edward Zalta. Available online at https://plato.stanford.edu/archives/spr2017/entries/computational-mind/

Rosenblueth, Arturo, Norbert Wiener, and Julian Bigelow, 'Behavior, Purpose and Teleology', *Philosophy of Science*, 10. 1(1943), 18–24.

Schweizer, Paul, 'Algorithms Implemented in Space and Time', *Selected Papers from the 50th Anniversary Convention of the AISB* (2014), 128–136 (London: The AISB).

Searle, John, *The Rediscovery of Mind* (Cambridge, MA: MIT Press, 1992).

Seth, Anil, 'The Cybernetic Bayesian Brain', in *Open MIND*, 35 (Frankfurt am Main: MIND Group, 2015).

Shannon, Claude, 'A Mathematical Theory of Information', *Bell System Technical Journal*, 27(1948), 379–423 and 623–656.

Sprevak, Mark, and Matteo Colombo, eds., *The Routledge Handbook of the Computational Mind* (London: Routledge, 2018).

Thompson, Evan, *Mind in Life* (Cambridge, MA: Harvard University Press, 2007).

Trappenberg, Thomas P., *Fundamentals of Cognitive Neuroscience* (Cambridge, MA: MIT Press, 2009).

Turing, Alan, 'On Computable Numbers, with an Application to the *Entscheidungsproblem*', *Proceedings of the London Mathematical Society*, 2. 42(1936), 230–265.

Varela, Francisco, Evan Thompson, and Eleanor Rosch, *The Embodied Mind* (Cambridge, MA: MIT Press, 1991).

Villalobos, Mario, *The Biological Roots of Cognition and the Social Origins of Mind* (Edinburgh: University of Edinburgh, 2015).

Villalobos, Mario, 'Nonequilibrium Thermodynamic Stability: The Apparent Teleology of Living Beings', in *Proceedings of the Artificial Life Conference 2016*, ed. by Carlos Gershenson, Tom Froese, Jesus M. Siqueiros, Wendy Aguilar, Eduardo J. Izquierdo and Hiroki Sayama (London: MIT Press, 2016), pp. 702–703.

Villalobos, Mario and Joe Dewhurst, 'Enactive Autonomy in Computational Systems', *Synthese*, 195. 5(2017a), 1891–1908.

Villalobos, Mario and Joe Dewhurst, 'Why Post-cognitivism Does Not (Necessarily) Entail Anti-computationalism', *Adaptive Behavior*, 25. 3(2017b), 117–128.

von Neumann, John, *First Draft of a Report on the EDVAC*, Contract No. W-670-ORD-4926 (Between the United States Army Ordnance Department and the University of Pennsylvania, 1945).

von Neumann, John, *The Computer and the Brain*, 2nd edn (New Haven and London: Yale University Press, 1952/2000).

Walter, W. Grey, 'An Electromechanical Animal', *Dialectica*, 4. 3(1950), 206–213.

Ward, Dave, David Silverman and Mario Villalobos, 'Introduction: The Varieties of Enactivism', *Topoi*, 36. 3(2017), 365–375.

Wiener, Norbert, *Cybernetics: Or Control and Communication in the Animal and the Machine* (Cambridge, MA: MIT Press, 1948).

Wiese, Wanja, and Thomas Metzinger, 'Vanilla PP for Philosophers: A Primer on Predictive Processing', *Philosophy and Predictive Processing* (2017), available at http s://predictive-mind.net/papers.

10 Towards a mutually beneficial integration of History and Philosophy of Science

The case of Jean Perrin

Klodian Coko

Introduction

Since the 1960s, there have been many efforts to defend the relevance of History of Science to Philosophy of Science, and *vice versa*. For the most part, these efforts have been limited to providing an abstract rationale for a closer integration between the two fields, as opposed to showing: (a) how such an integrated work is to be produced concretely, and (b) how an integrated approach can lead us to a better understanding of past and/or current science.[1] In this chapter, I argue that the most promising way to integrate the history and philosophy of science is the historicist-hermeneutic approach. I will present the main features of the historicist-hermeneutic approach, and will show, concretely, how it can provide a mutually beneficial integration of the History of Science and the Philosophy of Science. More specifically, I will employ the historicist-hermeneutic approach to elucidate one of the most problematic historical case studies in philosophy of science: namely, Jean Perrin's argument for molecular reality, which he formulated at the beginning of the twentieth century.

Jean Baptiste Perrin (1870–1942) was a French physicist who is widely credited by historians of science with providing a conclusive argument for the existence of atoms and molecules.[2] Perrin's argument ended the debates that took place during the long nineteenth century over the existence of these unobservable entities.[3] The most famous part of Perrin's argument was his description of thirteen different ways to determine Avogadro's number (N): the number of atoms, ions, and molecules contained in a gram-atom, gram-ion, and gram-mole of a substance, respectively. The different determinations included Perrin's own three, which were based on the experimental study of the height distribution, mean displacement, and mean rotation of Brownian particles.[4] For his experimental work in proving the discontinuous structure of matter, Perrin was awarded the Nobel Prize in Physics in 1926.

As well as being of interest to historians of science, Perrin's argument has also been the focus of much interest from philosophers of science. We can discern two, relatively independent, philosophical treatments of Perrin's argument. On the one hand, Perrin's argument is often presented as a case of *multiple determination*. Multiple determination is the epistemic strategy of using multiple, independent procedures to establish the same result.[5] It is widely regarded as a very important strategy by both working scientists and philosophers of science.[6] One contrived example that is used to illustrate the multiple determination strategy is that of independent witnesses: if several witnesses testify that an event occurred, and we can be certain that the witnesses' testimonies are independent, we can safely conclude that the event did occur. It would be an improbable coincidence for multiple witnesses, independently of one another, to fabricate exactly the same story. In the context of discussing the merits of multiple determination, Perrin's argument is presented as the paradigmatic case demonstrating the epistemic force of the strategy. When it comes to providing the specific grounds underlying Perrin's argument, however, not much is said besides – what we may call – the blunt rationale of multiple determination: namely, that it would have been a highly improbable coincidence for thirteen independent procedures to point at the same value for the number of molecules contained in a unit of substance, and yet for there not to be any such things as molecules. Not much analysis is provided regarding the role that multiple determination played in Perrin's experimental work and in convincing the scientific community. What makes matters more perplexing is that philosophers who have looked more closely at Perrin's argument have reached different (and often contradictory) conclusions, not only regarding the structure of Perrin's argument, but also regarding the role that the multiple determination of N played in it.[7]

Perrin's efforts were influenced by the late nineteenth-century recognition that an experiment in physics, in general, and the experimental investigation of unobservable entities in particular, required the use of complex instruments, experimental procedures, and the employment of theoretical and experimental auxiliary assumptions. Lacking direct observable evidence for the existence of atoms and molecules, Perrin's efforts were concentrated on what he considered to be the next best thing: the determination of the various magnitudes (mostly size and velocity) of the (hypothetical) molecules *via* independently theoretically-dependent routes.[8] The extremely remarkable agreement on the numerical values for the molecular magnitudes calculated by independently theoretically-dependent determinations gave rise to a strong no-coincidence argument, which was used to argue both for the correctness of the values determined, and the validity of the theoretical and experimental auxiliary assumptions underlying the different determinations.[9] The argument's structure, however, was more complex than the one encountered in the blunt rationale. There were structural elements of Perrin's argument which, although neglected in the various philosophical accounts, were responsible for its strength and, ultimately, for its success.

I argue that the integration of historical and philosophical perspectives provided by the historicist-hermeneutic approach is necessary for understanding the structure of Perrin's argument. This approach will also allow a conceptual framework to be developed that will allow us to understand the structure and the epistemic force of the multiple determination strategy. By following the historicist-hermeneutic approach, I emphasise both the historical context and the temporal development of Perrin's argument (as opposed to only looking at its final and finished form). I locate the main elements responsible for the argument's success. I argue that Perrin's argument was the result of his clear understanding of the philosophical and scientific challenges facing the empirical verification of claims regarding the existence of unobservable entities such as atoms and molecules at the beginning of the twentieth century.

The chapter is structured as follows. In section 10.1, I present some traditional approaches to HPS, which could be used to explore the Perrin case, and I indicate their shortcomings.[10] In the following sections, I employ the historicist hermeneutic approach to elucidate the reasoning underlying Perrin's work on Brownian movement and his argument for molecular reality demonstrating some of the advantages of this approach. In section 10.2, I present Perrin's early views on scientific methodology, which were very influential in shaping his experimental reasoning. In sections 10.3 and 10.4, I describe Perrin's early experimental work on the phenomenon of Brownian Movement. In section 10.5, I present the importance of this experimental work for his argument for molecular reality. In section 10.6, I show the epistemic import of the independent determinations of Avogadro's Number. In section 10.7, I present Perrin's later experimental work on Brownian movement and its importance for his argument for molecular reality. I conclude with some thoughts regarding the necessity of the historicist-hermeneutic approach for understanding the structure and epistemic import of the multiple determination strategy.[11]

10.1 Jean Perrin and traditional approaches to HPS

Perrin's argument for atomism was successful in ending nearly one hundred years of debates over the existence of unobservable atoms and molecules. It is not a surprise that it has been the focus of much philosophical interest. We can discern two, relatively independent, philosophical treatments. Firstly, Perrin's argument is often presented as a case of multiple determination. Philosophers of science who claim that the ability to establish the same result by means of independent procedures is an important epistemic strategy often refer to Perrin and to his thirteen determinations of Avogadro's number.[12] These philosophers claim that Perrin's reasoning was a no-coincidence argument. Namely, it would have been an improbable coincidence for thirteen different determination procedures to arrive at the same value for the number of molecules contained in a unit of substance, and for entities such as molecules not to exist. Other than the offering of this rationale, however, not much analysis is devoted to the actual role that the multiple determination of N played in Perrin's experimental work.

Secondly, Perrin's argument has been the subject of detailed case-studies by other philosophers of science. The proclaimed goal of these case-studies is to provide the exact reasoning behind Perrin's argument for molecular reality. The philosophers who have used Perrin's argument as a case-study have tried to appropriate it for their own purposes. Unsurprisingly, these philosophers have arrived at different, and often contradictory, conclusions, not only regarding the reasoning behind Perrin's argument, but also the role that the multiple determination of *N* played in it. For example, Perrin's argument has been interpreted as an 'inference to the best explanation' by Gilbert Harman (1965, p. 89). Clark Glymour (1975, p. 403), on the other hand, regarded Perrin's case as an instance of his account of 'bootstrapping confirmation'. Wesley Salmon (1984, pp. 213–226) claimed that Perrin's argument has the structure of a 'conjunctive common cause argument'. Nancy Cartwright (1983, p. 83), building on Salmon's 'common cause' interpretation, has argued that Perrin made an 'inference to the most probable cause'. Cartwright's interpretation was in turn challenged by Deborah Mayo. For Mayo (1986), Perrin's argument constituted a 'severe testing' for the molecular explanation of the phenomenon of Brownian movement. Peter Achinstein (2001, pp. 243–265) has claimed that his account of evidence offers the best interpretation of Perrin's reasoning. Achinstein's interpretation has been criticised by Stathis Psillos (2011) who argues that Perrin used Bayesian reasoning to provide a crucial experiment for the reality of atoms. Bas van Fraassen (2009) has argued that Perrin did not intend to, and thus did not establish, the real existence of atoms and molecules. He only provided 'empirical grounding' for one of the most important parameters of the kinetic theory: Avogadro's Number. Robert Hudson (2013, pp. 103–138) has claimed that Perrin's was not a case of multiple determination (or 'robustness', as he calls it), but a case of reliable process reasoning.

We are thus left with a number of different interpretations of both Perrin's reasoning for molecular reality and the role that the multiple determination of *N* played in it. It is not difficult to locate the source of the problem: all the above interpretations belong to a flawed way of integrating History and Philosophy of Science, namely what I call the illustrative model of HPS. Attention now will be turned to considering the weaknesses of this approach and other efforts to integrate the History of Science and the Philosophy of Science.

10.1.1 The illustrative model of HPS

All the interpretations of Perrin's argument mentioned above belong to the illustrative use of history (or 'illustrative model of HPS'). The illustrative model of HPS is explained in the following way: philosophers of science have pre-established conceptions – regarding, for instance, the nature of scientific explanation, the nature of the relationship between theory and evidence, the nature of theory – and they often use the historical material to illustrate or offer support for these pre-established conceptions. There is, therefore, the danger that philosophers might have misunderstood, or even intentionally distorted, the argumentative reasoning of the scientists they study. The following quote from Ernan McMullin (1970, p. 18) identifies the major shortcomings and pitfalls of the illustrative approach:

It makes use of the great scientists of the past as lay figures in what seems to be a historical analysis but really is not. They are manipulated to make a philosophical point which, however valid it may be in itself, was really not theirs, or at least is not really shown using the proper methods of the historian to have been theirs.

There have been several attempts to deal with the problems of the illustrative approach. In the next two subsections, I describe the two most influential approaches.

10.1.2 The confrontation model of HPS (the VPI programme)

One way to deal with the problems generated by the illustrative model of HPS is the so-called 'confrontation model of HPS'.[13] This approach was supported by philosopher of science Larry Laudan and his group at the Virginia Polytechnic Institute in the 1980s.[14] Their proposal was that instead of letting philosophical pre-conceptions influence the interpretations of historical facts, we should use the historical material to test philosophical theses, more or less in the same way that scientists use empirical evidence to test theoretical hypotheses. Although it was in fashion during the 1980s, this approach is now considered outdated.

10.1.3 The methodology of historiographical research programmes

Another way to address the problems with the illustrative use of history was proposed by philosopher of science Imre Lakatos (1970) and was put to fruition by his disciples. This approach is known as the 'Methodology of Historiographical Research Programmes'. It suggests scholars: (a) embrace the fact that historical material is always influenced by philosophical interpretations, (b) try to impose different philosophical interpretations on the historical material (a process known as 'rational reconstruction'), and (c) choose the philosophical interpretation which offers the most consistent account of the historical material. Although popular in the late 1970s and the 1980s, this approach was also eventually abandoned.

One can easily find in the literature a compendium of the problems with each one of these three approaches to HPS, as well as analyses of why they all lead straight to 'disaster'.[15] I will only mention what I think is the greatest weakness which they all have in common: the fact that there is nothing historical about them. The philosophical conceptions that are illustrated, tested, or used to rationally reconstruct the historical material, are presented as fixed and eternal, and as being applicable to all times and all places, rather than to a specific historical context. Therefore, these approaches fail to pay attention both to the historical context and to the historical development of knowledge.

10.1.4 The historicist-hermeneutic approach to iHPS

A more promising way to integrate historical and philosophical accounts of scientific practice is the 'Historicist-Hermeneutic Approach to iHPS'.[16] Aspects of this approach have been supported by various scholars during the second half of the twentieth century.[17] As its name indicates, this approach has two main parts: historicism and hermeneutics. The term 'historicism' has been used in many ways in the history of philosophy. In this context, it refers strictly to the very simple and intuitive, and yet often neglected, idea that the best way to understand something is to know how it came about. Philosopher of science Dudley Shapere (1977, p. 496) concisely described the historicist approach for understanding science:

> The question of why science today believes the peculiar things it does about the universe, and why it is willing to consider the alternatives it does, requires attention to the question of how science has come to think in those ways.

For Shapere, science is a self-sustainable and self-generating enterprise that has an intrinsically temporal dimension. If the philosopher's duty is to understand science, paying attention to this temporal dimension is essential. It is not enough to consider a 'slice' of scientific work at a moment in time.

The 'hermeneutic' part of the approach can also be understood by using another of Shapere's (1984, p. 185) dictums: 'We learn how to learn as we learn'. This dictum also reflects the idea that science is a self-transforming enterprise with no concepts and methodological precepts that are applicable to all times and all places. Supporters of the historicist-hermeneutic approach argue that this dictum applies also at the meta-level: to the philosophical efforts for understanding science. We are continuously bettering our philosophical understanding of science as we study science.

According to the historicist-hermeneutic approach, History of Science and Philosophy of Science are not separate endeavours, but activities that are already closely intertwined. The use of philosophical notions, tools, and concepts such as *theory, experiment, evidence, hypothesis, confirmation, multiple determination*, and so on, are necessary for understanding even a single episode in the history of science (this is the case even if this use of philosophy may go unnoticed to the historians themselves). The insight offered by the historicist-hermeneutic approach is that we can use this necessary intertwinement for the improvement of both philosophy of science and history of science. For instance, we can improve our understanding of philosophical notions such as the ones mentioned above, by studying concrete historical episodes where such notions are employed, and by studying their historical development through time. We can then use these improved and more precise philosophical notions to achieve a better understanding of other concrete episodes from past and/or present science, and then use these conclusions to achieve further elaboration and refinement of the said philosophical notions and their historical development, and so forth.[18]

In the rest of the chapter, I will use the historicist-hermeneutic approach to understand the reasoning behind Perrin's argument for molecular reality, which, as we saw earlier in this section, is one of the most problematic historical case studies in the philosophy of science. In doing so, I will shed light on the Perrin case and demonstrate both the use of, and the advantages of this approach. My approach is historicist because I will argue for the importance of the temporal development of Perrin's thought for understanding the reasoning underlying his argument, and especially for understanding the role that the independent determinations of *N* played in it. I argue, one important reason various philosophers have arrived at such different assessments of Perrin's argument is because their accounts are based on the final versions of Perrin's argument. Almost all the philosophical interpretations of Perrin's argument mentioned above are based on the English translations of Perrin's influential 1909 paper 'Brownian Movement and Molecular Reality' and, in addition, his 1913 book *The Atoms*.[19] From these accounts, philosophers cherry-pick the elements of Perrin's argument that are most beneficial to their philosophical positions, and disregard the elements that do not fit. In contrast, my approach is hermeneutic because I do not have, as a starting point, some preconceived philosophical position, which I will try to illustrate, test, or use to rationally reconstruct the historical material. My starting point is the blunt rationale for multiple determination, which, in the end, will be further clarified and elucidated by its 'friction' with the historical material.

10.2 Perrin's early methodological views

Jean Baptiste Perrin was born in Lille, on September 30, 1870. He was raised in Lyon, where he also received his early education. He moved to Paris to enter a class of special mathematics at the lycée Janson de Sailly. Studying under Émile Lacour, young Perrin was encouraged to prepare for the École Normale Supérieure. He entered the prestigious Parisian school in 1891. He was immediately attracted to experimental physics and studied under Marcel Brillouin, one of the few French scholars who supported the kinetic theory of gases at the time. In 1895, after refusing a teaching position in secondary education, Perrin was appointed *agrégé-préparateur* at the École Normale. At the same time, he began his experimental work, first on cathode, and then on Röntgen rays. This early experimental work aimed to provide experimental evidence for the existence of atoms.[20]

Perrin's earliest thoughts on scientific methodology can be found in his book *Traité de Chimie Physique: Les Principes*, published in 1903.[21] From the book's general approach it is clear that, for Perrin, what was at stake in the contemporary atomic debates was not simply the question of whether there was enough evidence to warrant belief in the existence of atoms and molecules, but even more fundamentally, whether it was necessary for physical science to postulate the existence of such unobservable entities, and what could count as confirmatory evidence for their existence.

In the book's preface, Perrin discussed the two fundamental methods of physical science. The first is the inductive method, which is characterised by a sure and slow march, from the recording of particular empirical facts, to the formulation of general principles. It is a method characterised by the defiance of all mystery and metaphysics and the disdain for everything that cannot be reduced to perceivable empirical facts.[22] Opposite the inductive method stands the deductive method, which mostly provides 'explanations of the visible by the invisible'. More specifically, the deductive method 'consists in imagining for matter a structure the direct perception of which still escapes our imperfect senses, and such that its knowledge would allow to deductively predict the visible properties of the universe.'[23] Contrary to the prevailing philosophical atmosphere of the *fin de siècle* in France, Perrin argued that, rather than being incompatible, the two methods could be fruitfully combined to investigate the properties of matter which escaped empirical detection. This could be achieved without abandoning the inductive principle that physical science is fundamentally based on empirical facts, and without retreating into metaphysics. One simply had to accept the intuitive idea that what is empirically detectable is not limited to what is currently detectable, but that it could be extended with the development of new methods and the invention of more advanced scientific instruments. For Perrin, the atomic-molecular hypotheses had proved their fruitfulness and legitimacy by being able to deductively predict a variety of facts, which were then empirically confirmed. What was still needed was direct empirical observation, which would transform these hypotheses into a confirmed reality.

Les Principes was a textbook aiming to present the fundamental principles of physical chemistry. It did not do much for expounding the existing evidence in favour of the molecular hypothesis. In an article he had published in 1901 in the *Revue Scientifique* – which was addressed to a wider audience – Perrin had presented this evidence in a detailed manner and had explained the nature of the support it provided for the molecular hypothesis.[24] Most of the evidential support came from the molecular hypotheses made in the context of the kinetic theory of gases.[25] These hypotheses not only offered explanations for the visible properties of gases and liquids, but also offered numerical approximations for various molecular magnitudes such as the velocity and the diameter of molecules, and Avogadro's number (N). Perrin recognised that, in itself, this was not a strong argument for the existence of molecules, or for the validity of the molecular values calculated.[26] There was nothing remarkable about the ability of the molecular hypotheses made in the context of the kinetic theory of gases to explain known facts or provide theoretical values for the molecular magnitudes: these hypotheses were constructed in the first place exactly in order to explain and accommodate the observable facts. The importance of these first numerical approximations was that they could be compared with values for the same magnitudes derived independently from the investigation of other phenomena. The atomic chemical theory, for example, was another hypothesis which was invoked to explain the empirical evidence from the chemical combination of substances. The explanations of the phenomena of electrolysis and of the newly discovered phenomena of cathode and X-rays had also given rise to yet other hypotheses regarding the discontinuous structure of matter – although at a level deeper than that of molecules. If one could derive similar values for the molecular magnitudes from the consideration of such diverse phenomena, then one could put forward a very strong argument for molecular reality, which was the next best thing in light of the lack of direct empirical observation.[27]

10.3 The phenomenon of Brownian movement and the qualitative triangulation of molecular reality: the molecular hypothesis as a logical induction

Beginning in 1901, Perrin became fully acquainted with the phenomenon of Brownian movement: the incessant and completely irregular movement of microscopic particles when suspended in liquids.[28] Although it was known for the most part of the nineteenth century, it was only during the 1870s that the importance of the phenomenon for the kinetic-molecular hypothesis was recognised.[29] Perrin's main source on the topic was the work of the French physicist Léon Gouy (1895) who, by the end of the nineteenth century, had experimentally established the basic properties of the phenomenon and had demonstrated its independence from all imaginable external influences.

From his earliest writings on the topic, Perrin argued that the phenomenon of Brownian movement offered a different kind of evidence for molecular reality from the *a priori* considerations made in the kinetic theory of gases.[30] Whereas in the kinetic theory one postulated *a priori* a hypothetical molecular structure for matter from which to *deduce* the observable facts, the phenomenon of Brownian movement moved in the opposite direction: it provided directly observable evidence that could be used to *inductively* infer a molecular structure for matter. It provided, what Perrin calls, a *logical induction*.[31]

In sum, Perrin's argument was the following: the basic characteristics of Brownian movement, as established by Gouy and other nineteenth-century investigators included the fact that it never stopped, complete irregularity, dependence on the size of the suspended particles, dependence on the temperature of the suspending liquid, independence from the nature of the particles, and independence from any external influences. These characteristics led naturally to the conclusion that the phenomenon was caused by the internal movements of the liquid itself. There is, therefore, a continuous movement of the internal parts of the liquid. The distribution of motion in a fluid does not de-coordinate indefinitely. Therefore, the liquid ought to be composed of elastic granules which are in permanent motion. If such granules have no existence, it is not apparent why there is a limit to the de-coordination of motion, and how a phenomenon such as that of Brownian movement is possible. For Perrin, the empirical examination of Brownian movement alone, independently of any kinetic considerations, was sufficient to logically suggest that every fluid is composed of elastic granules, animated by a perpetual motion.[32] Perrin concluded:

> Now, we only need to call these granules *molecules*, to recognize an old hypothesis, glimpsed by the intuition of Epicurus and Lucretius, revived and clarified by Bernoulli, and developed by Clausius and Maxwell. Only that, this hypothesis is no longer in our eyes *a priori*: it ranks as a logical induction, inspired from the observation of phenomena, in the same way that, for example, the undulatory theory of light is inspired, but not imposed, by the known properties of light.[33]

The phenomenon of Brownian movement thus provided an inductive argument for the existence of unobservable molecules, independent from the deductive argument provided by the kinetic theory of gases. The logical induction fell short of establishing the kinetic-molecular explanation of Brownian movement, because it did not prove that the discontinuous parts of the liquid causing the phenomenon were the same as the molecules postulated by the kinetic theory of gases. To establish this identity, what was needed was *an independent quantitative determination*: one would have to use the observable properties of Brownian movement to calculate the quantities of the magnitudes causing them, and then compare the results with the values for the molecular magnitudes provided by the kinetic theory of gases. This is exactly the experimental path that Perrin followed, beginning in 1908.

10.4 The quantitative triangulation of molecular magnitudes

Perrin argued that the best way to connect the observable characteristics of Brownian movement with the kinetic-molecular movements (supposedly) causing them, was to consider the suspended particles as giant molecules (for example, like molecules of sugar in a solution of sugar water).[34] To establish the identity of the granular movements causing the Brownian movement with the molecular movements postulated by the kinetic theory, one had to triangulate: one had to calculate the kinetic energy of a Brownian particle and compare it with the kinetic energy that the kinetic theory had deduced for an isolated molecule at the same temperature. The ingenuity of Perrin's experimental approach consisted in finding a way to calculate the kinetic energy of a Brownian particle that did not require the calculation of its velocity, which, by this time, it was realised it was impossible to measure.[35] Perrin developed the hypothesis that the Brownian particles of a homogeneous emulsion, because of their irregular movements, ought to distribute themselves in the same way as the (hypothetical) air molecules under the influence of gravity. It had been known since the eighteenth century that the density of a gas in equilibrium decreases with altitude according to an exponential law.[36] Perrin's idea was that, if he could establish that the height distribution of Brownian particles obeyed the exponential law, he would have confirmed the hypothesis that the gas laws extended to Brownian particles and, thus, have in the behaviour of the suspended Brownian particles a magnification, in a visible scale, of the behaviour of the unobservable (and hypothetical) molecules. In early 1908, Perrin conducted his famous *height distribution experiments*, which established that the height distribution of Brownian particles was indeed exponential.[37] Further, Perrin claimed that he could explain this exponential height distribution in a way that allowed the determination of the osmotic pressure (k) of a single Brownian particle. Perrin devised his height distribution equation, which allowed the calculation of k, if one could determine: the mass of a Brownian particle in a homogeneous emulsion (m), the density of the Brownian particle (p), and the ratio of the concentration of Brownian particles at two different levels of the emulsion $\left(\frac{n_0}{n}\right)$.

$$2,3 log\frac{n_0}{n} = \tfrac{1}{k}mgh\left(1 - \tfrac{1}{p}\right) \text{ (Logarithm to base 10)}$$

Perrin attempted the experimental determination of the osmotic pressure of a Brownian particle, in early 1908 and published his results in the *Comptes Rendus de l'Académie des Sciences*. His aim was to prove 'that molecular agitation is an actual cause, and unique cause, of Brownian movement.'[38] Leaving aside the ingenious ways that Perrin invented to circumvent all the difficulties that surrounded the experimental calculation of the magnitudes appearing in the height distribution equation and, making a long story short, he found the osmotic pressure (k) exerted by a Brownian particle to be equal to 360.10^{-16}.[39] Perrin compared the osmotic pressure exerted by n Brownian particles with the pressure that, according to the kinetic theory, was exerted by n molecules of a gas. This pressure would be equal to $n\frac{RT}{N}$ (with R being the constant of perfect gases, T the absolute temperature, and N the number of molecules contained in one gram-molecule – which theoretical considerations from the viscosity of gases placed it around 7.10^{23}). After making all the calculations Perrin found that the pressure exerted by n molecules of a gas was equal to $n\times343.10^{-16}$; almost equal to the osmotic pressure of n Brownian particles, assuming the validity of the value for N.[40] The conclusion of Perrin's first experimental paper from 1908, reads:

> *The mean kinetic energy of a colloid granule is therefore equal to that of a molecule* ... At the same time, the kinetic theory of fluids seems a little more fortified, and the molecules a little more tangible. Their number N in a gram-mole, deduced from the previous equality, assumed to be correct, is $6,7.10^{23}$.[41]

10.5 The structure and epistemic import of Perrin's height distribution experiments

In section 10.1, I showed how different philosophers, starting from different pre-conceived philosophical positions, have offered different interpretations of Perrin's experimental work. In sections 10.2 to 10.4, by following the historicist approach, I placed Perrin's experiments on the height distribution of Brownian particles in their temporal dimension. Having done so, we are now able to tackle the important question: What was the structure and epistemic import of Perrin's height distribution experiments? The historicist approach shows a number of things about Perrin's height distribution experiments:

1   They were part of a case of multiple determination (or triangulation), and not simply the confirmation of a theoretical prediction made by the kinetic theory of gases. Perrin's experimental work was a continuation of his early methodological views. The height distribution experiments offered an independent, experimental (or inductive) determination of k and, subsequently, of values for other molecular magnitudes which were first determined in a deductive manner in the kinetic theory of gases. The precise quantitative agreement between the two determinations established that the molecules, hypothesised in the kinetic theory of gases, had a real existence and that they were identical with the 'granules' that, based on inductive reasoning from empirical observations, ought to be the cause of the phenomenon of Brownian movement.[42]

2   The two determinations of k were theoretically independent. They were based on different reasoning processes (inductive *vs.* deductive), on the consideration of different phenomena (Brownian movement *vs.* viscosity of gases) and, most importantly, on theoretically independent auxiliary assumptions.

3   The numerical agreement achieved was very striking, especially if one considered the possible numerical values for k that could be the result of the height distribution experiments. According to Perrin, the range of possible experimental values extended from zero to infinity.[43]

4   Perrin used the agreement between the numerical value for k obtained in his height distribution experiments with the value for k theoretically deduced in the kinetic theory of gases to argue, not only about the validity of the result, but – perhaps even more importantly – about the validity of the central theoretical and experimental auxiliary assumptions upon which his height distribution experiments were based.[44] The theoretical auxiliary assumptions included: the theorem of the equipartition of energy (which was central for the claim that Brownian particles behaved just like the molecules postulated by the kinetic theory of gases), the claim that the laws of perfect gases extended to uniform emulsions (with the particles of a uniform dilute emulsion behaving like the

molecules of a gas or liquid in equilibrium), the extension of Stokes' law to the order of magnitude of Brownian movement, and the claim that molecular movement was the (unique) cause of Brownian movement. The numerical concordance was also used to argue about the validity of the experimental methods employed to measure the magnitudes that appeared in the height distribution equation. These included the methods used: to prepare a uniform emulsion with spherical granules of equal diameter, to calculate the mass of the granules, and to determine the ratio $\frac{n_0}{n}$. Perrin's underlying reasoning was that it would be a remarkable coincidence for independent determination procedures to arrive at almost identical numerical values for the values of the molecular magnitudes measured, and yet for the (theoretical and experimental) auxiliary assumptions underlying them to be essentially flawed.[45] This form of reasoning is crucial. Because of the large number and the precarious nature of the auxiliary assumptions required to determine the magnitudes appearing in the height distribution equation, no theoretical or experimental determination, *by itself*, could ever be sufficient to establish both the validity of the result and the validity of the determination procedure. Only the strong no-coincidence argument that emerged when independently theoretically-dependent procedures converged on the same value for k, could be used to argue both for the validity of the result and the validity of the determination procedures.

5   Perrin thought that one important feature of the height distribution experiments, which distinguished it from other efforts of determining the molecular magnitudes, was that they allowed an unlimited precision in determining the values for k and N.[46] As Perrin would often repeat in his writings, providing a precise value for N from the height distribution experiments was simply a question of conducting very careful experiments and making precise calculations of the magnitudes appearing in the height distribution equation.[47]

10.6 The epistemic import of the independent determinations of Avogadro's number

To recapitulate, after conducting the height distribution experiments, Perrin claimed that the numerical concordance he had achieved: (a) established without a doubt the kinetic-molecular explanation of Brownian movement, (b) justified his theoretical and experimental approach, and (c) provided a first determination of the various molecular magnitudes. It is only after these initial experiments that providing a precise value for Avogadro's number (N) became central to Perrin's experimental work.[48] Avogadro's number, because of its direct connection with other molecular magnitudes, could serve as a sort of common ground for coordinating between the determinations of the various molecular magnitudes coming from the consideration of different phenomena.[49] At the time when Perrin concluded his height distribution experiments (end of 1908), besides the value for Avogadro's number calculated from the theoretical considerations made in the context of the kinetic theory of gases, there emerged four other determinations. All of them (largely) agreed with Perrin's. Calculations of the charge of the electron (e), which were conducted at the Cavendish laboratory, placed N between 43.10^{22} and 96.10^{22}. Max Planck's and H.A. Lorentz's calculations, which were based on the theory of black-body radiation, gave for N the values 61.10^{22} and 77.10^{22}, respectively. Finally, Ernest Rutherford's calculations of e, which were based on the study of radioactivity, placed N between 62.10^{22} and 77.10^{22}. To these values, Perrin added his own determination by 'a method which seems to me *direct and susceptible to an unlimited precision.*'[50] By October 1908, Perrin had conducted three series of experiments with Brownian particles of different sizes. They involved calculations for 13000 particles and 16000 readings. Despite the variation of the different parameters, all the experiment series gave – within the limit of experimental error – the same invariant value for N: $70,5.10^{22}$.

In section 10.1, I showed how the various philosophers who used Perrin's work as a case study disagree not only regarding the structure of his argument for molecular reality, but also regarding the role that the independent determinations of N played in it. Again, the historicist approach allows us to tackle this issue conclusively. To recapitulate, the question we are faced with is this: what is the importance of these additional concordant determinations of N for Perrin? This is especially given that (as demonstrated in the previous section) he had claimed that his experimental work had already established both the kinetic-molecular explanation of Brownian movement and the value of N. By reading Perrin's writing from this period, we can infer several reasons why he thought these additional concordant determinations were important.

1   *Lack of Discordance.* Given that Perrin had already established the validity of his experimental approach and was certain that it could be used to provide precise values for N, we could say that the fact that it was not contradicted by the other determinations was a huge relief. A discordant result could have raised doubts about the molecular theory of Brownian movement and required further experimental investigation to determine the source of disagreement. In fact, this is exactly what happened when the first experimental efforts to verify Einstein's mathematical theory on the molecular origin of Brownian movement led to results that were discordant with Perrin's (see section 10.7 below). The lack of discordance was extremely striking if one considered the *a priori* possibilities for the values for Avogadro's number that were possible in each one of the different determinations.

2   *The Variety of the Phenomena Considered.* The different determinations were based on the theoretical consideration or the experimental investigation of different phenomena: viscosity of gases, Brownian movement,

black body radiation, radioactivity, and the electric charge of ions.[51] This variety of phenomena not only provided the required theoretical independence of the different determinations, but was to be expected (or even required) given that one was trying to determine the value of a fundamental magnitude concerning the 'building blocks' of observable phenomena.

3   *The Genetic Independence of the Determinations.* Besides being theoretically independent, the different determinations of N were also genetically independent.[52] What is meant by genetic independence is simply the fact that the different determinations were conducted independently of one another. Although Perrin did not use the term, he was fully aware of the possible objection that the achieved agreement could be construed as a case of experimental calibration or mutual adjustment of the experimental results. Perrin explicitly stated how lucky he was that most of the experimental determinations of N were conducted concurrently with his, without the different researchers having knowledge of each other's results. This precluded the possibility that they had (deliberately or even subconsciously) calibrated their results to achieve agreement.[53]

4   *The Offering of Mutual Support.* The theoretical independence of the determinations, the genetic independence of determinations, the number of determinations, the quantitative nature of the agreement, the variety of phenomena considered, the lack of discordant results, were elements which were used to construe a very strong no-coincidence argument to support the validity of the auxiliary assumptions underlying the different determinations: especially the determinations which were based on the investigation of new phenomena (like black-body radiation and radioactivity) and which were thought to be based on more speculative and untested auxiliary assumptions. The rationale of this no-coincidence argument was that it would be an improbable coincidence for the various determinations to arrive at the same value for Avogadro's number, and yet for the auxiliary assumptions underlying them to be essentially flawed.

10.7 The emergence of discordance: mathematical theories of Brownian movement

Perrin was able to provide two additional, concordant values for N via his experimental study of the mean horizontal displacement and mean rotation of Brownian particles, respectively. What did Perrin think was the importance of these additional determinations of N? Continuing with the historicist approach and looking at the events from a temporal perspective – as opposed to looking only at the final form of Perrin's argument – shows that Perrin's aim in conducting this additional experimental research was not to offer another determination of N, but to use the concordance in order to remove the doubts regarding the molecular-kinetic explanation of Brownian movement that had emerged when the first efforts to experimentally verify Albert Einstein's mathematical work on Brownian motion failed to do so.

In 1905, without even knowing that the phenomenon of Brownian movement had been already observed and studied for around eighty years, Einstein produced a mathematical formula which described the average horizontal displacement that, according to the kinetic theory of heat, the (hypothetical) molecular movement ought to be causing on microscopic particles suspended in a liquid:

$$\lambda_x = \sqrt{t}\sqrt{\frac{RT}{N}\frac{1}{3\pi kP}}$$

where, λ_x is the mean horizontal displacement of a suspended (Brownian) particle, t is the time interval during which the displacement is measured, T is the absolute temperature, N is Avogadro's number, k is the coefficient of viscosity of the liquid, and P the radius of the particle.[54]

The important thing about Einstein's formula was that it defined the horizontal displacement of a suspended particle without involving its real velocity, which could not be calculated because of the extremely complicated path the particle ought to be describing during a specific time interval. This opened the way for an experimental confirmation, given that all the magnitudes could (theoretically, at least) be experimentally determined. After presenting this equation, Einstein concluded his 1905 paper by hoping 'that some enquirer may succeed shortly in solving the problem suggested here.'[55]

The publication of Einstein's theoretical work on Brownian motion was followed by three independent verification attempts. They were by The Svedberg in Sweden, Max Seddig in Germany, and Victor Henri in France. They all, and independently of one another, failed to verify the formula. Svedberg argued that his experimental results offered a rough verification of the formula, but his claims were rejected by his contemporaries, including Perrin and Einstein.[56] Seddig accepted the failure, but blamed his experimental method.[57] The verification failure that came from the quantitative, cinematographic study of Brownian movement, undertaken by Victor Henri at the Collège de France, in the beginning of 1908, was the failure that had the most impact on Perrin and on the community of French physicists at the time. Contrary to Svedberg and Seddig, Henri concluded that Einstein's displacement formula did not apply to the Brownian movement of the particles he had experimentally studied.[58] Henri's results were interpreted as a failure to establish the kinetic-molecular movement as the unique cause of the phenomenon of Brownian movement. They were consistent with the position, still defended by many French physicists at the time, that the electric actions exerted from the ions of the liquid on electrically charged suspended particles were an additional cause of the phenomenon.[59]

Perrin believed that the height distribution experiments had established beyond any reasonable doubt the kinetic-molecular explanation of Brownian movement. The failure to experimentally verify Einstein's displacement formula put in front of him the choice between the inexactness of molecular explanation and the inexactness of the formula. Perrin chose the latter option, believing that some unjustified assumption had entered into Einstein's reasoning. Nevertheless, after suggestions made by Aimé Cotton and Paul Langevin, he attempted a verification of the displacement formula by using Brownian particles of exactly known radius which he had used in his height distribution experiments.[60]

Perrin conducted the first measurements with the help of his doctoral students.[61] Surprisingly, the initial displacement measurements offered a satisfactory agreement with the value for N calculated in the height distribution measurements of the same particles. In 1909, Perrin announced a mean value for N, calculated by around 3000 displacement recordings, equal to $70,5.10^{22}$. This value was identical with the value for N determined in the height distribution experiments and remained relatively invariant to changes of the various experimental parameters. Perrin used the numerical agreement to support the validity of both the experimental procedures employed to determine the magnitudes appearing in Einstein's formula and the theoretical assumptions underlying Einstein's mathematical derivation.[62]

After verifying Einstein's displacement formula, Perrin saw the possibility of an experimental test of Einstein's equation for the rotational Brownian movement. Einstein had theoretically demonstrated that the molecular impacts, besides a translational movement, imparted on the suspended microscopic particles also a rotational movement. At the basis of Einstein's equation of mean rotation was the equipartition of energy theorem, which claimed that, at the same temperature, the mean kinetic energy of rotation of a suspended Brownian particle was equal to its mean kinetic energy of translation, and both equal to the mean kinetic energy of an isolated molecule (and all this independently of the size of the granule). Perrin's stated aim behind this experimental effort was not another confirmation of the molecular theory of Brownian movement, or another determination of N, but the confirmation of the theoretical assumptions underlying Einstein's rotation equation:[63] in particular, the invariance of the equipartition of energy theorem to changes of the various parameters (especially to changes in the size of Brownian particles).[64]

Conclusion: Towards a two-way, mutually beneficial, integration of History and Philosophy of Science

In this chapter, I have argued for the necessity of the historicist-hermeneutic approach for achieving a mutually beneficial integration of History of Science and Philosophy of Science. As outlined above, aspects of the historicist-hermeneutic approach have been supported by various scholars during the last fifty years. I demonstrated how this approach can be applied concretely to solve one of the most problematic case-studies in philosophy of science: the reasoning underlying Jean Perrin's argument for molecular reality. I have argued that Perrin's was a case of multiple determination. Perrin put forward a no-coincidence argument for the existence of molecules, which was based on the agreement between multiple, independent determinations of Avogadro's number (and consequently, other molecular magnitudes). The blunt rationale of the argument was the following: it would be a highly improbable coincidence for multiple, independent determinations of molecular magnitudes to achieve concordant results, and yet for there not to be any molecules. The careful application of the historicist-hermeneutic approach, however, shows that there were additional structural elements of Perrin's argument that were responsible for its exceptional strength and, ultimately, for its success. They were the following:

1 The argument was based on a *quantitative* multiple determination. That is, the independent determinations concerned specific numerical values of the molecular magnitudes.
2 There was a close agreement between the independent determinations. This agreement became even more striking, if one considered the possible values for the molecular magnitudes that could have been the result of each one of the determinations.
3 There was a (relatively) large number of determinations which converged on the same result.
4 The different determinations were theoretically independent: that is, they were based on independent theoretical assumptions.
5 The different determinations were genetically independent and no effort was made to mutually adjust the numerical values calculated by theoretically independent procedures.
6 The different determinations were based on the investigation of unrelated phenomena.
7 The high quality and reliability of some of the determinations.
8 There was not even one discordant result, despite the large number of determinations.
9 When objections and discordant results which challenged Perrin's determination of molecular magnitudes emerged, Perrin conclusively resolved the discordance.

Following the historicist-hermeneutic approach it is possible to develop a conceptual framework for dealing with the structure and epistemic importance of the multiple determination strategy in scientific practice. The historicist-hermeneutic approach, as employed in Perrin's case, shows the existence of several structural elements upon which the strength of the no-coincidence argument – the defining feature of the multiple determination strategy – depends. These elements are:

1　The number of determinations: the more the determinations that produce the same result, the stronger the no-coincidence argument.
2　The theoretical independence of the determinations: the more theoretically independent the determination procedures that establish the same result are, the stronger the no-coincidence argument.
3　The genetic independence of the determinations: the more genetically independent the determinations procedures that establish the same result are, the stronger the no-coincidence argument.
4　The reliability of the determinations: the more reliable the determination procedures that establish the same result are, the stronger the no-coincidence argument.
5　The quality (or clarity) of the result established by independent determinations: the clearer or more precise is the result upon which the independent determinations agree, the stronger the no-coincidence argument.
6　The quality of the convergence: the more the determination procedures are judged to have established the same result, the stronger the no-coincidence argument.
7　The complexity of the independently established result: the more complex is the result that is established by independent determinations, the stronger the no-coincidence argument.
8　The existence of discordant results and/or conflict with accepted knowledge: the less the discordant results and/or the conflict with accepted knowledge, the stronger the no-coincidence argument.

Continuing with the historicist-hermeneutic approach, we can use this preliminary conceptual framework to understand and evaluate the epistemic force of other cases of multiple determination, from past or current science. For example, we can use it to understand why in some cases of multiple determination the no-coincidence argument succeeds, whereas in other cases it fails. The implementation of this step will demonstrate the relevance of Philosophy of Science to History of Science. This step is different from the traditional use of philosophical pre-conceptions to interpret the historical material. And this is because from the interaction of this initial conceptual framework with the historical material it is possible to further sharpen and elucidate our initial framework. This could be done, for example, by noticing other structural elements that influence the strength of the no-coincidence argument underlying the multiple determination strategy. The implementation of this step will demonstrate the relevance of History of Science to Philosophy of Science. We can use this more developed framework to elucidate and evaluate other (or even the same) cases of multiple determination.[65] And so forth. Our efforts to understand science in its historical dimension are themselves open-ended.

Notes

1 Jutta Schickore (2011) offers a historical account of these efforts.
2 Klodian Coko (2015a), pp. 71–72.
3 Rocke (1984), Nye (1972) and Chalmers (2009) offer comprehensible and accessible historical accounts of the nineteenth century atomism debates.
4 Jean Perrin (1909c, 1913).
5 This very same strategy is referred to by different terms in the literature. Some of them are: 'robustness', 'independent confirmation', 'triangulation'. I use the term 'multiple determination', because it is more transparent and less technical than the other terms. We should always keep in mind, however, that this term is used here only to refer to a particular epistemic strategy. The analysis of the structure and epistemic import of this strategy offered here is not in any way dependent on the term used to refer to it.
6 William Wimsatt (1981); Ian Hacking (1981); Allan Franklin (1986); Nancy Cartwright (1991); Sylvia Culp (1994), to mention only a few.
7 The structure and epistemic import of Perrin's argument has not been a topic of interest to historians of science, who are mostly limited to providing descriptive accounts of Perrin's experimental work.
8 I use the term 'independently theoretically-dependent' to refer to the fact that each one of the determination procedures that were used to determine Avogadro's Number was dependent on theoretical auxiliary assumptions, but the set of theoretical assumptions underlying a specific procedure was largely independent from the set of theoretical assumptions underlying another procedure. That is to say, the validity of the theoretical assumptions underlying one procedure was largely independent from the validity of the theoretical assumptions underlying another procedure.
9 A no-coincidence argument for the validity of an experimental result is the argument that the only explanation that would not make the result an improbable coincidence, is the explanation that the result is valid.
10 I will use the acronym HPS (History and Philosophy of Science) to refer to the traditional approaches, and the acronym iHPS (Integrated History and Philosophy of Science) to refer to the historicist-hermeneutic approach. There are two main reasons for this. First, the use of the acronym iHPS to refer to the traditional approaches would be anachronistic. The term 'iHPS' was coined only recently, namely, in 2007, in the first international conference of Integrated HPS, at the University of Pittsburgh. Secondly, and most importantly, the acronym was coined exactly to differentiate Integrated HPS, an approach that is both historical and philosophical, from traditional approaches to HPS.
11 I use the term 'epistemic force of the multiple determination strategy' to denote the strength of the argument that multiple determination provides for the validity of the claims that are established by using the strategy.
12 Ian Hacking (1983), pp. 186–209; Peter Kosso (1988); Sylvia Culp (1995); James Woodward (2006), and many others.
13 Schickore (2011), p. 456.
14 Larry Laudan et al. (1986).
15 Thomas Nickles (1986). Hans Radder (1997) and Schickore (2011) provide accessible and detailed accounts of these problems.
16 Schickore (2011) introduces the term and provides an account of this approach.
17 Ernan McMullin (1974); Dudley Shapere (1977); Richard M. Burian (1977); Thomas Nickles (1995); Schickore (2011); Hasok Chang (2012).
18 Schickore (2011), pp. 471–474.
19 Psillos (2011) is a notable exception.
20 Jean Perrin (1897).

21 Jean Perrin (1903).
22 Ibid., p. vii.
23 Ibid. All translations are mine.
24 Jean Perrin (1901).
25 Ibid., pp. 451–455.
26 Ibid., p. 449.
27 Ibid., pp. 456–459.
28 Jean Perrin (1923), pp. 22–28.
29 Mary J. Nye (1972), chapter 1.
30 Jean Perrin (1906).
31 Ibid., p. 338.
32 Ibid., pp. 335–336.
33 Ibid., p. 338.
34 Jean Perrin (1905) p. 58.
35 Nye (1972), p. 101.
36 Jean Perrin (1905), p. 60.
37 Jean Perrin, (1908a).
38 Ibid., p. 968.
39 Ibid., p. 969.
40 Ibid.
41 Ibid., p. 970. Italics in the original.
42 Perrin (1908d), pp. 514–515.
43 Jean Perrin (1908b), p. 531.
44 Perrin (1908d), p. 528.
45 Viewed in this manner, Perrin's experimental work seems to provide a direct response to Pierre Duhem's famous critique of experimental method from the 1890s. This way of reasoning is not necessarily troublesome or vicious. Since the emergence of the problem of the experimenters' regress – which I regard as a more general version of the *Duhem thesis* – philosophers have been aware of the inter-dependency existing between the correctness of an experimental result and the reliability of the experimental procedure used to establish that result: a correct experimental result is generally considered to be one produced with a reliable experimental procedure; but a reliable experimental procedure is the one that produces the correct result (Collins 1985). In other words, we don't know if we have obtained the correct result unless we have used a reliable procedure, but we don't know if we have used a reliable procedure unless we have obtained the correct result. Insofar as the no-coincidence argument from multiple determination helps to break the regress by arguing about the validity of the result, it can also be used to argue about the reliability of the procedures.
46 Ibid., pp. 529–532.
47 Ibid.; Jean Perrin (1909a; 1913).
48 Perrin (1908b), p. 532.
49 Perrin, Jean (1908c), p. 594.
50 Ibid., p. 595. Italics in the original.
51 Perrin (1909c), pp. 93–113.
52 See Soler (2012) for a discussion.
53 Perrin (1909c), p. 108.
54 Albert Einstein (1956a) p. 18.
55 Ibid., p. 18.
56 Perrin (1909c), p. 74; Albert Einstein (1956b).
57 Nye (1972), p. 125.
58 Victor Henri (1908), p.1026.
59 Aimé Cotton (1908), p. 739; Jacques Duclaux (1908).
60 Perrin (1909a), p. 32.

61 Chaudesaigues (1908),; Perrin and Dabrowski (1909).
62 Chaudesaigues (1908), p. 1045; Perrin (1909a), p. 33.
63 Jean Perrin (1909b), p. 550.
64 Perrin (1909c), p. 92.
65 Klodian Coko, (2015b).

Bibliography

Achinstein, Peter, *The Book of Evidence* (Oxford: Oxford University Press, 2001).
Burian, Richard M., 'More than A Marriage of Convenience: On the Inextricability of History and Philosophy of Science', *Philosophy of Science*, 44.1(1997), 1–42.
Cartwright, Nancy, *How the Laws of Physics Lie* (Oxford: Oxford University Press, 1983).
Cartwright, Nancy, 'Replicability, Reproducibility and Robustness: Comments on Harry Collins', *History of Political Economy*, 23(1991),143–155.
Chalmers, Alan, *The Scientist's Atom and the Philosopher's Stone* (Dordrecht: Springer, 2009).
Chang, Hasok, 'Beyond Case Studies: History as Philosophy', in *Integrating History and Philosophy of Science, Boston Studies in the Philosophy of Science 263*, ed. by Seymour Mauskopf and Tad Schmaltz (Dordrecht: Springer, 2012), pp. 109–124.
Chaudesaigues, M., 'Le mouvement brownien et le formule d'Einstein', *Comptes Rendus*, 147(1908),1044–1046.
Coko, Klodian, 'Epistemology of a Believing Historian: Making Sense of Duhem's Anti-atomism', *Studies in History and Philosophy of Science*, 50(2015a), 71–82.
Coko, Klodian, The Structure and Epistemic Import of Multiple Determination in Empirical Science, Dissertation Thesis, Indiana University, 2015b.
Collins, Harry M., *Changing Order: Replication and Induction in Scientific Practice* (London: SAGE Publications, 1985).
Cotton, Aimé, 'Recherches récentes sur les mouvements browniens', *La Revue du Mois*, 5(1908),737–741.
Culp, Sylvia, 'Defending Robustness: The Bacterial Mesosome as a Test Case', *PSA: Proceedings of the Biennial Meeting of the Philosophy of Science Association*, 1 (1994),46–57.
Culp, Sylvia, 'Objectivity in Experimental Inquiry: Breaking Data-Technique Circles', *Philosophy of Science*, 62.3(1995), 438–458.
Duclaux, Jacques, 'Pression osmotique et mouvement brownien', *Comptes Rendus*, 147 (1908),131–134.
Einstein, Albert, 'On the Movement of Small Particles Suspended in a Stationary Liquid Demanded by the Molecular-Kinetic Theory of Heat', in *Investigations on the Theory of the Brownian Movement by Albert Einstein*, ed. by R. Fürth, transl. by A.D. Cowper (New York: Dover Publications, 1956), pp. 1–18.
Einstein, Albert, 'Theoretical Observation on the Brownian Motion', in *Investigations on the Theory of the Brownian Movement by Albert Einstein*, ed. by R. Fürth, transl. by A.D. Cowper, (New York: Dover Publications, 1956), pp. 63–67.
Franklin, Allan, *The Neglect of Experiment* (Cambridge: Cambridge University Press, 1986).
Glymour, Clark, 'Relevant Evidence', *The Journal of Philosophy*, 72.14(1975), 403–426.
Gouy, Léon, 'Le mouvement brownien et les mouvements moléculaires', *Revue Générale des Sciences*, 6(1895),1–7.

Hacking, Ian, 'Do We See Through a Microscope?', *Pacific Philosophical Quarterly*, 63(1981),305–322.

Hacking, Ian, *Representing and Intervening: Introductory Topics in the Philosophy of Natural Science* (Cambridge: Cambridge University Press, 1983).

Harman, Gilbert H. 'The Inference to the Best Explanation', *The Philosophical Review*, 74.1(1965), 88–95.

Henri, Victor, 'Étude cinématographique des mouvement brownien', *Comptes Rendus*, 146(1908),1024–1026.

Hudson, Robert G., *Seeing Things: The Philosophy of Reliable Observation* (Oxford: Oxford University Press, 2013).

Kosso, Peter, 'Dimensions of Observability', *The British Journal for the Philosophy of Science*, 39.4(1988), 449–467.

Lakatos, Imre, 'History of Science and its Rational Reconstructions', *PSA: Proceedings of the Biennial Meeting of the Philosophy of Science Association* (1970), 91–136.

Laudan, Larry, Arthur Donovan, Rachel Laudan, Peter Barker, Harold Brown, Jarrett Leplin, Paul Thagard and Steve Wykstra, 'Scientific Change: Philosophical Models and Historical Research', *Synthese*, 79.2(1986), 141–223.

Mayo, Deborah G., 'Cartwright, Causality, and Coincidence', *PSA: Proceedings of the Biennial Meeting of the Philosophy of Science Association*, 1(1986),42–58.

McMullin, Ernan, 'The History and Philosophy of Science: A Taxonomy', in *Minnesota Studies in the Philosophy of Science*, ed. by R.H. Stuewer (Minneapolis: University of Minnesota Press, 1970), pp. 12–67.

McMullin, Ernan, 'History and Philosophy of Science: A Marriage of Convenience?', *PSA: Proceedings of the Biennial Meeting of the Philosophy of Science Association* (1974), 585–601.

Nickles, Thomas, 'Remarks on the Use of History as Evidence', *Synthese*, 69.2(1986), 253–266.

Nickles, Thomas, 'Philosophy of Science and History of Science', *Osiris*, 2nd Series, 10 (1995),138–163.

Nye, Mary J., *Molecular Reality: A Perspective on the Scientific Work of Jean Perrin* (New York: American Elsevier Company, 1972).

Perrin, Jean, *Rayons Cathodiques et Rayons de Röntgen: Étude Expérimentale* (Paris: Gauthier-Villars et fils, 1897).

Perrin, Jean, 'Les hypothèses moléculaires', *Revue Scientifique*, 15.15(1901), 449–461.

Perrin, Jean, *Traité de Chimie Physique: Les Principes* (Paris: Gauthier-Villars, 1903).

Perrin, Jean, 'Mécanisme de l'électrisation de contact et solutions colloïdales', *Journal de Chimie Physique*, 3(1905),50–110.

Perrin, Jean, 'La discontinuité de la matière', *Revue de Mois*, 1(1906),323–344.

Perrin, Jean, 'L'agitation moléculaire et le mouvement brownien', *Comptes Rendus*, 147(1908a), 967–970.

Perrin, Jean, 'L'origine du mouvement brownien', *Comptes Rendus*, 147(1908b) 530–532.

Perrin, Jean, 'Grandeur des molécules et charge de l'électron', *Comptes Rendus*, 147 (1908),594–596.

Perrin, Jean, 'Peut-on peser un atome avec précision?', *La Revue du Mois*, 6(1908c), 513–538.

Perrin, Jean, 'Mouvement brownien et molécules', *Journal de Physique Théorique et Appliquée*, 9(1909a), 5–39.

Perrin, Jean, 'Le mouvement brownien de rotation', *Comptes Rendus*, 149(1909b), 549–551.

Perrin, Jean, 'Mouvement brownien et réalité moléculaire', *Annales de Chimie et de Physique*, 18(1909c), 1–114.

Perrin, Jean, *Les Atomes* (Paris: Libraire Félix Alcan, 1913).

Perrin, Jean, *Notice sur les Travaux Scientifiques de M. Jean Perrin* (Toulouse: Édouard Privat, 1923).

Perrin, Jean and M. Dabrowski, 'Mouvement brownien et constants moléculaires', *Comptes Rendus*, 149(1909),477–479.

Psillos, Stathis, 'Making Contact with Molecules: On Achinstein and Perrin', in *Philosophy of Science Matters: The Philosophy of Peter Achinstein*, ed. by Gregory J. Morgan (Oxford: Oxford University Press, 2011), pp. 177–190.

Radder, Hans, 'Philosophy and History of Science: Beyond the Kuhnian Paradigm', *Studies in History and Philosophy of Science*, 28.4(1997), 633–655.

Rocke, Alan, *Atomism in the Nineteenth Century: From Dalton to Cannizzaro* (Ohio: Ohio University Press, 1984).

Salmon, Wesley, *Scientific Explanation and the Causal Structure of the World* (Princeton: Princeton University Press, 1984).

Schickore, Jutta, 'More Thoughts on HPS: Another 20 Years Later', *Perspectives on Science*, 19.4(2011), 453–481.

Shapere, Dudley, 'What Can the Theory of Knowledge Learn from the History of Knowledge?', *The Monist*, 60.4(1977), 488–508.

Shapere, Dudley, *Reason and the Search for Knowledge, Boston Studies in the Philosophy of Science*, 78 (Dordrecht: D. Riedel Publishing Company, 1984).

Soler, Léna, 'Robustness of Results and Robustness of Derivations: The Internal Architecture of a Solid Experimental Proof' in *Characterizing the Robustness of Science: After the Practice Turn in Philosophy of Science*, ed. by Léna Soler *et al.*, Boston Studies in the Philosophy of Science, 292 (Dordrecht: Springer, 2012), pp. 227–266.

van Fraassen, Bas C., 'The Perils of Perrin, in the Hands of Philosophers', *Philosophical Studies*, 143(2009),5–24.

Wimsatt, William, 'Robustness, Reliability, and Overdetermination', in *Scientific Inquiry and the Social Sciences*, ed. by M.B. Brewer and B.E. Collins (San Francisco: Jossey-Bass, 1981), pp. 124–163.

Woodward, James, 'Some Varieties of Robustness', *Journal of Economic Methodology*, 13.2(2006), 219–240.

11 Revitalising a nineteenth century debate about life (which has been done to death)

Or, how to live with historiographical pluralism

Alex Aylward

Introduction

Reflecting on the relationship between History of Science (HS) and Philosophy of Science (PS), most in our field would admit that the latter can – perhaps *should* – draw fruitful lessons from the former. Philosophers want their theories about science to be borne out empirically, and to give a good account of the actual practice of science. Engagement with HS can help achieve these aims.[1] Perhaps less obvious is the value of PS for HS. As prominent iHPS scholar Theodore Arabatzis (2017) put it: 'What's in it for the Historian of Science?' His answer begins with Norwood Russell Hanson's observation that historical studies of science must grapple with metascientific concepts. Arabatzis (2017, p. 69) urges that 'philosophical reflection on those concepts can be (and, indeed, has been) historiographically fruitful.' He discusses the examples of 'epistemic values', 'experimentation', 'scientific discovery', and 'conceptual change', showing in each case how philosophical insights upon these matters can be (and have been) profitably deployed in historical scholarship. Scientific practices, past and present, often involve issues of philosophical weight, engagement with which can be invaluable in gaining thorough historical understanding. What I want to suggest is that PS can also inform HS externally to the specifics of the 'metascientific concepts' encountered during any particular HS project, at the more removed level of deciding which episodes to study historically, which methodologies to employ in doing so, and what we hope to gain from our investigations. In particular, I will show how recent work in PS on pluralism and perspectivism can enhance HS. Section 11.1 sets out this approach in the abstract, whilst sections 11.2 and 11.3 apply it to a case-study from HS: a debate over the nature of life which took place at London's Royal College of Surgeons in the early nineteenth century. Reflecting on the results of this case-study, I end by summarising the ways in which applying lessons from pluralism and perspectivism in PS can benefit our historiographical practices.

11.1 Pluralising historiographical perspectives: a lesson for HS from PS

When we possess multiple differing accounts of a historical episode, we often see them as *competing*. Philosopher Katherina Kinzel asks in her 2016 essay how we might go about *restricting* such historiographical pluralism; which criteria should we apply in judging between competing accounts and how do we apply them? She concludes that, though most inadequate accounts may be fairly straightforwardly disposed with, we often lack strong enough criteria to neutrally adjudge a single 'best' account. Consequently, 'we will have to live with some degree of pluralism in historiography'.[2] The heuristic guiding Kinzel's rigorous study – one which I will be contesting – is that historiographic pluralism is bad. Its presence indicates that we have not yet arrived at the one best account of an episode; or if we have, that we nevertheless have been unable to demonstrate the inadequacy of its competitors. There is an analogy here with the sciences. The existence of alternative scientific accounts of the same phenomenon has tended to lead scientists and philosophers alike to ask which theory is 'best', and should thus be maintained at the expense of 'competitors'. The enduring centrality of the problem of 'theory choice' in PS is indicative of the value traditionally attached to the pursuit of monism in the face of pluralism. Increasingly, however, philosophers of science are urging that we turn a critical eye on our traditional commitment to monism, and even that we *actively cultivate* pluralism in science.[3]

One variety of scientific pluralism which can be of substantial value to HS, I suggest, is rooted in scientific perspectivism, which holds that a scientific theory or model only ever provides a partial account of its target phenomenon.[4] Thus, scientists are only granted *a perspective* upon their object of study, among many which are possible. Competing scientific accounts will often provide alternative perspectives upon the phenomenon of interest, and will emphasise and illuminate (or obfuscate) different aspects of the target phenomenon to varying degrees. Once the partiality of scientific perspectives is acknowledged, monism appears a rather limiting virtue; a plurality of perspectives is required for anything approaching a 'complete' account of the phenomenon of interest.[5] As perspectivist philosopher Michela Massimi (2018, p. 165) has recently put it, '[t]here cannot be an objective, unique, true description of the way the world is as soon as we acknowledge that our scientific knowledge is always from a specific vantage point'. Historians of science should heed these lessons from PS, for, as David Hull (1992, p. 472) noted, '[h]istory of science cannot be written from no perspective whatsoever'. The methodologies we adopt, and the historiographical frameworks we employ, influence which historical questions, and answers, we deem interesting and illuminating. Like science, HS is perspectival. Pursuing historiographical monism in the face of pluralism, then, appears misguided. We cannot hope for 'an objective, unique, true description' of any episode in the History of Science; hence, we stand to gain – as do scientists – from pluralising our perspectives. To be sure, historiographical pluralities already exist: else there would be little motivation for Kinzel's project.[6] Contra Kinzel, I am suggesting we embrace and learn from existing historiographical pluralism, and actively cultivate more.

Others before now have encouraged historiographical pluralism. Historian of science Robert Fox (2006, p. 412), for example, has written that, '[t]he more options we have as historians, the better. For different questions call for different methodological tools, and we need as broad a repertoire as possible.' Yet, there is something about my plea for pluralising *perspectives* that differentiates it from calls to proliferate our 'methodological tools' or similar. We employ tools; we have perspectives *on* things. The latter and not the former implies the constancy of our target phenomenon, even as we pluralise perspectives. It is unclear whether Fox's 'different questions' relate to the same historical episode or different ones. Rather than merely suggesting that we practice in our field a variety of historiographical methodologies (a hopefully uncontroversial claim), the lesson historians of science should draw from scientific perspectivism is that we should focus many different methodological tools *upon a particular case-study.*[7] Only by pluralising our historiographical perspectives *upon some particular episode* do we stand to gain a fuller account of that episode, as well as reflexive insights concerning the adopted perspectives themselves. The resultant plurality of accounts is not something we should aim to remedy à la Kinzel, but cultivate and learn from in ways I shall detail below.

Sometimes we may wish to apply this methodology of pluralising perspectives *de novo*, to a past scientific episode that is yet to be studied. Certainly, this is the surest way of gaining a 'complete' account of the episode. However, here I am concerned with revisiting episodes *already* explored from one historiographical perspective, and consciously *re*-exploring them from an alternative one. The potential benefits of revisiting particular episodes in this way are many, including (but not limited to): the testing of historiographical/philosophical frameworks against known examples from the history of science (think of it as testing new kinds of microscope against existing kinds working on different principles); highlighting problems with our existing historiographical categories; alerting us to interesting historical and/or philosophical questions not obvious from existing perspectives; exposing shortcomings of, or errors made from, existing perspectives.

The best way to be clear on what is meant by each of these benefits is through example. In section 11.2, I give an overview of my chosen case-study for re-exploration: the Abernethy–Lawrence controversy – a heated dispute in the 1810s over how best to explain vital phenomena, between two colleagues at London's Royal College of Surgeons (RCS). This case serves my aims well, because the various accounts of it written last century – there are several, hence my titular 'almost done to death' phrase – represent, I suggest, but one perspective upon the episode. After a brief characterisation of this perspective, I proceed in section 11.3 to revisit the affair from an alternative one, namely by utilising a historiographical framework suggested by historian of science and medicine Andrew Cunningham for thinking about the disciplines of anatomy and physiology in my period of interest. Though imperfect as applied to the Abernethy–Lawrence case, pursuing this perspective provides various benefits of the kinds set out above.

11.2 The Abernethy–Lawrence controversy

In 1799 William Lawrence (1783–1867), a promising sixteen-year-old apprentice, came under the patronage of esteemed surgeon John Abernethy (1764–1831). As was customary, Lawrence lived in his mentor's household for the duration of his apprenticeship, after which he was appointed anatomical demonstrator at St. Bartholomew's Hospital in 1801, and a member of the RCS in 1804. It was here that, in 1814, Abernethy delivered a set of introductory anatomical lectures entitled *An Enquiry into the Probability and Rationality of Mr. Hunter's Theory of Life.*[8] Extraordinary in scope, Abernethy's lectures tackled one of the most enticing problems of all: what is life?

According to Abernethy, what distinguished living from non-living matter was that, in the tradition of eighteenth-century 'Newtonian' ether-theories, the living was pervaded and animated by a subtle, immaterial, vital spirit.[9] Drawing liberally upon contemporary investigations of electrical and chemical phenomena, Abernethy mobilised this notion to explain a wide range of the so-called 'vital phenomena' displayed by living organisms, such as animation, sensibility and irritability. Lawrence objected, vocally and publicly, to his former mentor's account. Though names went unmentioned, Lawrence's (1816) RCS lectures – *An Introduction to Comparative Anatomy and Physiology* – represent a rebuttal of Abernethy, ridiculing his invocation of a subtle life-giving principle. Lawrence, inspired by French anatomist Xavier Bichat, advocated instead a variety of vitalism which emphasised 'organisation', and located vital functions in particular tissues.[10] Abernethy (1817) doubled-down on his position in his *Physiological Lectures*, which Lawrence (1819a) responded to in turn in his *Lectures on Physiology, Zoology, and the Natural History of Man.*[11]

The personal nature of the dispute was not lost upon commentators in the specialist and general press. The conservative *Quarterly Review* blasted Lawrence for 'converting the lecture-room of the College into a school of materialism', whilst outrage was felt at his 'most coarse and virulent invective against his former patron'.[12] Lawrence's defenders in the press held that moral and theological convictions should be set aside in discussing scientific matters. A failure to do so, one commentator reminded their readers, had left Galileo 'imprisoned in a dungeon for truths afterwards confirmed by Newton'.[13] In an effort to rescue his reputation, Lawrence withdrew his lectures from circulation (though they were pirated by various publishers after he lost the copyright). Lawrence's retreat from the controversy had the desired effect, as he went on to enjoy a glittering medical career, culminating in his appointment in 1858 as Serjeant Surgeon to Queen Victoria.

Intensive social-historical study of the episode began in the 1960s by June Goodfield-Toulmin (1966, 1969) and Owsei Temkin (1963), and continued two decades later by Stephen Jacyna (1983b) and Adrian Desmond (1987, 1989). Together, these authors successfully situated this pre-Victorian debate on the science of life within the context of religious, political, and class tensions in radicalising Britain. Roughly, the resultant *perspectival* account has it that Abernethy's principle-vitalism was aligned with the conservative political establishment, orthodox Christianity, and patriotism, whilst the Francophile Lawrence's doctrines reflected the politically dangerous, materialistic atheism emanating from the continent in the wake of the French Revolution. Elucidated in this way the Abernethy–Lawrence saga, according to Jacyna (1983b, p. 311), 'merely constituted a fortissimo statement of a recurrent motif'.

This account of the affair is now standard, being cited regularly as an exemplary case-study of the relation between early nineteenth century British science and the political upheavals of the time.[14] Yet the above is but one possible perspective on the Abernethy–Lawrence affair. Further, there are some concrete considerations which confirm our PS-informed suspicion that this perspective (like any) provides a less-than-complete account of the episode. Take, for instance, historian of biology Karl Figlio's (1976, p. 20) criticisms of Goodfield-Toulmin's and Temkin's accounts. Figlio cites evidence of French resistance to materialism, British opposition to immaterial vital principles, and the 'deep appreciation of French thought in Scottish philosophical circles', in suggesting that Goodfield-Toulmin and Temkin overemphasised national differences (pp. 33–38). Any one perspective is indeed liable to over- or under-emphasise certain aspects of the phenomenon under study.

Additional pressure can be applied to the received perspective by interrogating the supposed alignment of Abernethy's subtle-fluid-based physiology with his conservative politics and mainstream Christianity. Firstly, ether and God need not be inextricably linked, as John Christie (1981, p. 94) has shown for the eighteenth century Edinburgh physician William Cullen, whose chemical ether *replaced* God as the effective causal agent in nature. Secondly, Abernethy's invocation of electricity in discussing the vital fluid – on which, more later – puts pressure on the characterisation of his doctrine as politically 'conservative', in contrast with Lawrence's 'radical' views. On electricity in this period, Iwan Rhys Morus (2011, pp. 32–38) argues that, whilst the 'plaything of fashionables', the application of galvanic and electrical studies to life and the body carried materialistic and 'radical' undertones (as Mary Shelley brought to bear in her 1818 novel *Frankenstein*). At worst, one could argue that the tight bundling of political persuasion, religious beliefs, and scientific views which is central to the received perspective is at risk of unravelling. At best, one must concede that the broad social-contextualisation perspective I have summarised is – like all perspectives – only partial.

11.3 A new perspective: the pen and the sword

My crude summary of the works of Goodfield-Toulmin, Temkin, Jacyna and Desmond represents but one possible perspective on the Abernethy–Lawrence affair, albeit an illuminating one. These authors attempted – successfully – to situate this dispute at the RCS within a historiographical framework emphasising a nexus of national, political, and religious tensions at the turn of the nineteenth century. Yet, as I have suggested, there are likely alternative perspectives from which we can give an account of the episode, which will be fruitful in different ways.

The alternative perspective I will pursue, based on work by historian Andrew Cunningham, takes different categories as historically significant, compared with the received perspective.[15] Ultimately, it will be clear that the new perspective is far from perfect.[16] No matter: no perspective is. This was a benefit of pluralising perspectives that I listed in the introduction: the testing of historiographical frameworks against known examples. In spite (and partly because of) its limitations, the perspective pursued here is still productive in the various ways suggested: it highlights problems with existing historiographical categories; alerts us to new and interesting historical and/or philosophical questions not obvious from other perspectives; exposes errors made from other perspectives; and, most straightforwardly, renders our account of this episode from the history of science more 'complete'.

The perspective takes the nature and role of certain *scientific disciplines* as its historiographical category of interest. In the early 2000s, historian of science and medicine Andrew Cunningham (2002, 2003, 2010) set out in a pair of articles – and later at book-length – a framework for 'recovering the disciplinary identity of physiology and anatomy before 1800'. The decades after 1800 were, for Cunningham, a watershed moment during which the disciplines of anatomy and physiology were transformed beyond recognition, taking on roughly the identity they retain to this day. Before this radical transformation, anatomy was the great experimental science of life, being a much richer and more active pursuit than we now conceive it to be. Physiology in this period was *purely theoretical*; the 'old-physiologist' was a *philosopher*, who weaved together

> the anatomical and other evidence he had acquired from experiment, observation and reading, and *reasoned* his way to understanding how it all functioned together in life [...] He took the anatomical *facts* and on them built his physiological *speculations*.[17]

To capture the relations between 'old-anatomy' and 'old-physiology', Cunningham re-deploys the adage of the pen and the sword. The philosophising physiologist required only the former, and allowing himself the rhetorical flourish, Cunningham bestows the latter upon the knife-wielding anatomist. These disciplinary identities were transformed out of recognition, Cunningham explains, by the emergence in the early to mid-nineteenth century of 'experimental physiology', in the mould of pioneering Frenchmen François Magendie (1783–1855) and Jean Pierre Flourens (1794–1867). Far from the philosophical interpretation of facts provided by anatomists, physiology of the Magendie School was itself an active, experimental, interventionist discipline. Experimental physiology replaced 'old-anatomy' as the great experimental science of life, the major difference being that it intervened not in the dead, but the living body.[18] Anatomy, meanwhile, became demonstrative, losing its active, interventionist, and experimental identity.

The Abernethy–Lawrence saga was already underway by the time Magendie penned the first textbook of his new experimental physiology in 1816, which would take some time to exert an appreciable influence in the generally Francophobic British context.[19] The timing of the affair, then, places it just within the remit of Cunningham's proposed framework of the nature of and relationship between the disciplines of anatomy and physiology. Both men drew liberally upon both disciplines in explicating their accounts of vitality, and as such, the doctrines they espoused in their lectures provide a tantalising test-case for Cunningham's framework, which is itself lacking in applications to concrete episodes.

Adopting Cunningham's framework as an alternative historiographical perspective, I will show that a good deal of Abernethy and Lawrence's disagreement over the nature of life can be made sense of in terms of disparities between each man's notions of good physiological practice. Different disciplines are (and were) defined by different, sometimes conflicting, sets of methods, aims and approaches, that encapsulate, shape and constrain the research undertaken under their banners. A physiologist might pose questions or solutions that a comparative anatomist would not dream of, or regard as important, or even coherent. Through adopting Cunningham's discipline-centred perspective, we will see that the *attitudes* our protagonists held concerning the disciplinary scopes of anatomy and physiology, and particularly their relationship to one another, elucidates a great deal of the disparity in their respective conceptions of life.

The case is not as simple as one man 'doing old-anatomy' and the other 'doing old-physiology'; both men wielded pen and sword. Despite this, Cunningham's characterisation of these disciplines appears promising, as we read Lawrence (1819a, p. 8) stating that comparative anatomy, 'furnishes the data, which constitute the basis of general physiology, of which the object is to determine the laws that regulate the phenomena exhibited by organized beings'. Abernethy also held that knowledge of vital processes must be based upon facts ascertained by comparative anatomy. He instructed his audience that to gain knowledge of vital processes in the exemplary manner of the late, revered surgeon John Hunter, 'it is necessary to refer to the facts contained in his Museum'; in other words, examination of the *structures* of various animal forms is a prerequisite to theorising about the actions and functions of animate beings and their parts.[20]

Both parties in the dispute, and indeed most of the wider scientific community, subscribed to the view that particular functions were localised in particular bodily structures.[21] Lawrence's commitment to this methodological heuristic is clear in his 1819 entry in *Rees's Cyclopaedia*, on the topic of 'monsters', or, individual organisms, 'in whom the body in general, or some large and conspicuous part of it, deviates remarkably from the accustomed formation'.[22] Such instances presented the researcher unrivalled opportunities:

Monsters, in which considerable parts are wanting, seem peculiarly likely
to assist in the prosecution of physiological researches. If we never saw
animals, except in a perfect state, we could not form just ideas of the
comparative importance of the different organs.[23]

Lawrence (1819b, pp. 7–8), eyes peeled for information pertaining to the
localisation of functions, recounts the tale of a particular 'monster' patient:

> In one case, where [...] an imperfect cerebrum seemed to exist, the child
> lived six days. The child was perfectly formed, excepting the head, and of
> usual size. It took no food, and had no evacuation. Respiration went on
> naturally: it did not cry, but often made a hideous whining noise [...] No
> signs of voluntary motions appeared, and the mother had less feeling of
> the child in utero, than in her former pregnancy.

From such a case, Lawrence explained, one may deduce that the coordination
of functions including nutrition, excretion and voluntary motion – but
excluding respiration, vocalisation, and growth – are localised in the portion
of the brain disrupted.

Abernethy, unlike Lawrence, invoked an immaterial principle of vitality per-
vading the organism. However, this was seemingly no deterrent to believing that
vital functions could be mapped to specific structures and organs. Indeed,
Abernethy (1814, p. 46) explained that, '[a]s what is deemed the complexity of
animal life increases, we find distinct organs allotted for each of these functions'.
He later instructs his audience that 'it is generally believed that all sensation is in
the brain, and that all volition proceeds from that organ'. He mobilised various
empirical results supporting the conclusion that the brain is the seat of volition,
including 'that the perceptions and intellect of animals increase in proportion as
the brain becomes larger and more complex' (p. 65), and the observation that,
'[if] a certain degree of pressure be made upon the brain, both feeling and
voluntary motion cease whilst it continues and return when it is removed' (p. 67).
Epitomising Abernethy's conception of anatomical knowledge, and its standing
relative to the high theorising of physiology, he starts from the basis of localising
volition in the brain, and argues henceforth for the existence of a subtle, vital
fluid (pp. 67–68):

> If then it be admitted that sensation exists in the brain, and that volition
> proceeds from that organ, it necessarily follows that motions must be
> transmitted to and fro along the nervous chords, whenever they take
> place [...] Physiologists were therefore led to conjecture that the nervous
> fibrils were tubular, and that they contained a subtle fluid, by means of
> which such motions were transmitted.

Both men, then, preached and practised localisation in research. However, only
in Lawrence's case was the resultant anatomical knowledge invoked to ground

his physiology on a strongly observable and empirical bedrock. For Abernethy, anatomical knowledge was a springboard for conjectural hypothesising.

For Lawrence, the facts of comparative anatomy made clear and obvious the conclusion that life relies upon organisation, even in the somewhat controversial case of mental processes. If it was not the brain that performed our mental functions, but rather an immaterial principle attached to or housed within it, then why, Lawrence mused, is the former so large and complex? If the brain itself is redundant with respect to mental processes, then the fact it is 'better fed, clothed, and lodged than any other part, and has less to do' is quite inexplicable.[24] Moreover, the tight correlation of mental powers and cerebral size and complexity throughout the 'great chain of being' represented, for Lawrence, *anatomical facts* demonstrating the importance of organisation to vital functions. True, most vital functions are present *throughout* the living kingdom, in creatures of disparate organisation. However, Lawrence emphasised that these *properties* of life were manifested to degrees and levels of perfection that varied just as widely, and importantly, in a manner that *correlated with* organisational gradations. The 'bare facts' of anatomy, were employed by both Abernethy and Lawrence; by the former to downplay the centrality of organisation, and by the latter to uphold it.

Localisation was a tool of Cunningham's 'old-anatomist'. Lawrence rarely departed from the programme of general anatomy forwarded by his idol, Xavier Bichat, who desired what Jacyna (1990, p. 162) has described as 'a topographical or natural-historical account of tissues to which was subjoined an analysis of their vital properties'. Lawrence (1819a, pp. 81–82), like Bichat, was content to rest his doctrine on a level that he saw as, at least presently, irreducible:

> To say that irritability is a property of living muscular fibres, is merely equivalent to the assertion, that such fibres have in all cases possessed the power of contraction. What then is the cause of irritability? I do not know, and cannot conjecture.

Lawrence professed a strong disinclination to going beyond anatomical facts; he might be read as an ultra-empiricist 'old-physiologist'. His physiology was 'shackled' by anatomy. The divergence between his and Abernethy's approach is clear, and concerned the extent of each man's empiricism; how far they were willing to conjecture, to *physiologise*, beyond anatomical facts. For the most part, Lawrence merely anatomised; he localised vital functions, classifying tissues by their status as the seat of a particular vital property. His physiology did not go much further, nor did he desire it to. For Abernethy, however, a physiology overly shackled by anatomy simply did not tell us enough of interest about the workings of organisms.

Because of this divergence, Lawrence and Abernethy reached an impasse when seeking to elucidate the *causes* of vital phenomena. Lawrence entertained a 'constant-conjunction' view of causation.[25] He wrote in his *Introduction to Comparative Anatomy and Physiology* that:

Experience can only exhibit the order and rule of succession of the phenomena, which indicate the action of the cause. When one event is
observed constantly to precede another, the first of these is called cause,
and the latter effect; and we believe that the preceding event has a power
of producing that which succeeds; although, in reality, we know only the
fact of succession.[26]

With such a humble epistemology of causation in place, there was little
opportunity for Lawrence's physiological speculations to roam too far from
anatomical matters-of-fact. But what was for Lawrence a safeguard against
unfounded conjecture represented for Abernethy a restrictive shackle upon
knowledge-making. Abernethy (1817, p. 97) despaired:

> If ... [Lawrence and other such thinkers] mean to insinuate, that we have
> no knowledge of cause or effect beyond that which results from mere
> observation, they publish at the same time, a libel on the human under-
> standing; a prohibition to rational enquiry, and a most severe satire, on
> themselves.

The 'common-sense' Scottish philosophers to whom Lawrence owed so much
were also at this time beginning to recognise within science a constructive role
for analogy and hypothesis.[27] Abernethy (1814, p. 8) embraced this development firmly: '[F]ormation of an hypothesis excites us to enquiries, which may
either confirm or confute our conjectures; and which may, by enabling us to
discover the deficient facts, convert our hypothesis into a theory'.

Indeed, Abernethy (1814, p. 13) thought it 'highly probable that it was
[Hunter's] hypothesis respecting life which incited him to enquiries by which
he has been able to supply the deficient facts, so as to establish his conjectures, or convert his hypothesis into a theory'. Hypothesising, for Abernethy, was a justified movement beyond the facts; only through conjecture can
we drive our researches forward and learn new things from enquiries we
might not otherwise have considered pursuing. Lawrence (1816, p. 177) was
not wholly opposed to hypotheses, but only thought them warranted when
'they are adduced with the array of philosophical deduction, because they
involve suppositions without any ground in observation or experience, the
only sources of our information on these subjects'. Lawrence's attitude left
open the possibility that vital spirits could be employed as a heuristic principle for guiding research, though he rightly interpreted Abernethy as wanting
his vital principle to do much more.[28]

The elder surgeon's hypothesising was analogy-driven, and it provoked
some of Lawrence's most devastating retorts. Abernethy (1814, p. 42) habi-
tually compared his vital principle with the mysterious force of electricity,
also purported to operate via a subtle and mobile fluid, urging his audience
that, '[t]he phænomena of life and electricity correspond'. Choosing a vital
property that the Swiss anatomist Albrecht von Haller had localised in

muscular tissue, irritability, for comparison with the electrical force, Abernethy (1814, p. 48) argued that '[t]he motions of electricity are characterized by their celerity and force; so are the motions of irritability. The motions of electricity are vibratory; so likewise are those of irritability'. Abernethy took great heart from Humphry Davy's recent work in electrochemistry, believing that the great chemist had 'solved the great and long hidden mystery of chemical attraction', showing its dependence upon 'the electric properties which the atoms of different species of matter possess' (p. 48). The bearing of these conclusions upon Abernethy's case was indirect. He believed they showed that electricity, this subtle and powerful principle, was pervasive in nature, 'and that it enters into the composition of everything, inanimate or animate' (p. 49). Hence, he reasoned, electricity or something *similar*, 'pervades organized bodies' (p. 51) and produces the vital processes within them, as electricity underpins the chemical changes undergone by inanimate matter. Analogy, for Abernethy, pointed to the probability and rationality of his theory of life.

Lawrence (1816, p. 169) ridiculed his adversary's approach: 'To make the matter more intelligible, this vital principle is compared to magnetism, to electricity, and to galvanism; or it is roundly stated to be oxygen. 'Tis like a camel, or like a whale, or like what you please'. These analogies, Lawrence maintained, did not enlighten: the nature of electricity was as mysterious as that of any purported vital principle. Moreover, the analogies proposed were without foundation. For Lawrence (1816, p. 171), '[i]dentity or similarity of cause can only be inferred from identity or resemblance of effect', but vital processes like digestion, growth, and sensibility differ vastly from the effects of the electrical force. Abernethy's analogical flourishes represented moves which Lawrence's philosophy simply did not permit; going beyond the observable anatomical facts. According to the received perspective, Lawrence's distaste for Abernethy's vital principle had its foundations in morals and politics. It is true, one of his objections to a 'vital principle' was its supposed affirmation of a transcendental power controlling human freedom; it was intended to 'impose a restraint upon vice stronger than Bow street or the Old Bailey can apply'.[29] However, it is equally clear that the invocation of a subtle, immaterial agent of vitality violated many of the rules bounding what was, for Lawrence, proper *physiologising*.

Taking Cunningham's lead, I have asked what light the roles of anatomy and physiology can shed on the Abernethy–Lawrence affair. We have seen that our two authors' conceptions of the scope of anatomical and physiological practices significantly diverged, in ways which illuminate the discrepancy in their respective accounts of life. The works of these men differ in the degree to which they exhibit what Cunningham suggests was the relationship between the 'old' styles of anatomy and physiology. Abernethy's propensity to hypothesise and analogise meant he deviated from the ideal of a physiology based exclusively upon *anatomical fact*. Abernethy was willing to go further beyond the brute facts than Lawrence's strict empiricism would allow. The Cunningham-inspired perspective pursued here has led us to consider issues neglected by the received social-contextualisation perspective, as well as telling us useful things about Cunningham's framework itself. The final section discusses these various fruits of pluralising our perspectives upon the Abernethy–Lawrence case.

11.4 A productive pluralism

At the close of section 11.2, we saw Karl Figlio accuse the received perspective of overemphasising national differences, with Abernethy and Lawrence painted as Francophobe and Francophile, respectively. I also suggested that certain aspects of Abernethy's principle-vitalism cause problems for interpreting his doctrine as consonant with conservative politics and orthodox Christianity. The upshot was that social, political, and religious considerations – upon which the received perspective focuses – are alone insufficient to fully capture Abernethy and Lawrence's disagreement over the nature of life. This is unsurprising, I suggest, because HS – like the science it investigates – is perspectival. Recent movements in PS acknowledge the perspectival nature of our knowledge-making practices, and thus recommend we pluralise our perspectives. We should do the same in HS. The resultant historiographical pluralism differs from existing calls for pluralism in HS; it is not simply a matter of tolerating and maintaining various methodological approaches within our discipline, but rather actively and consciously training a variety of such approaches *upon particular episodes*, in order to yield multiple perspectives. Besides gaining greater 'completeness' in our historical understanding of an episode, the potential benefits of my approach are several, including: (a) problematising our existing historiographical categories; (b) alerting us to new and interesting questions not obvious from other perspectives; (c) exposing errors or mischaracterisations made by other perspectives. Additionally, there is the reflexive benefit that, (d) via application to known cases in the history of science, we can develop and improve the adopted framework itself. By way of concluding this chapter, I will show how each benefit is manifested in our perspectival re-exploration of the Abernethy–Lawrence affair.

Our chosen perspective has trained our attention upon the ways in which each man thought about the proper scope, methods, and aims of the disciplines they practised, as well as the relationship between them. In turn, this has led us to explore the competing epistemologies and notions of causation at play in the debate. It turns out that a coherent, illuminating account of the disagreement between these practitioners can be offered at this level. To tell such a story does not, however, invalidate those already told at the level of the political, the social, and the religious. Rather than competing, we can view these different perspectives as complementary. The fact that Abernethy and Lawrence held differing epistemological commitments, including the proper relations they perceived to exist between anatomical and physiological practices, does not at all suggest that their political and religious differences were inconsequential. Indeed, further investigation may reveal interesting interrelationships between these two sets of considerations (Benefit b). Confessedly, Cunningham's (2010, p. xxi) pen-and-sword framework purposefully paints the state of 'old-anatomy' and 'old-physiology' in a 'somewhat static way', in order to emphasise the contrast with what came after. Thus, Cunningham underplays the heterogeneity within these disciplines, and in the relations between them. The case of Abernethy and Lawrence – figures united geographically, temporally and institutionally, but who nevertheless negotiated the relations of anatomy and physiology quite differently – can add some welcome nuance of a quite subtle kind to Cunningham's framework (Benefit d).

The perspectival account I have offered also highlights certain errors or mis-characterisations in the received accounts (Benefit c). As we saw, Lawrence sought to localise vital functions in particular tissues, and go no further; upon the causes of the functions, he could 'not conjecture'. He elaborated at length his views upon the properties of living and non-living matter, specifying that the former is 'governed by physical laws, such as attraction, gravitation, chemical affinity; and it exhibits physical properties, such as cohesion, elasticity, divisibility, &c. Living matter also exhibits these properties, and is subject in great measure to physical laws'. So far, so standard from a man accused by his contemporaries of physicalist-materialism. But he continues: 'But living bodies are endowed moreover with *a set of properties altogether different from these, and contrasting with them very remarkably*'.[30]

Indeed, Lawrence frequently derided crude attempts by some physiologists to reduce vital processes to mathematics and the physical sciences: 'One esti-mated the force of the heart as equal to 180,000lbs.; another reduced it to 8oz.; and both these conclusions are deduced from reasonings *clothed in all the imposing forms of the exact sciences*.'[31]

It is unsurprising, given the socio-political milieu explored in the received perspective, to find Lawrence's views misrepresented by politically-motivated conservative quarterlies. More recent commentators, though, are also guilty of misreading Lawrence. Adrian Desmond (1989, p. 117), for instance, explains that Lawrence 'believed that the ordinary laws of physics and chemistry were quite adequate to explain this life-giving organization'. This mischaracterisation is puzzling in light of the quotations given just above, in which Lawrence expli-citly details his anti-reductionism concerning vital phenomena. But such are the subtle hazards that come with Desmond's ambitious perspective, seeking as it does to contextualise thinkers within the broad political, social and religious tensions of their time. If Abernethy's political conservatism and patriotism are wedded to his vitalism, Lawrence's perceived radicalism and Francophilia might push us too far in interpreting his doctrines as reductionist and materialist. The perspective explored in this chapter, which encourages close engagement with the philosophies and methodologies at play, guards against the kind of error Des-mond makes (though surely leaves us exposed to many others).

Given our augmented understanding of Lawrence's position, we can be confident that, by any measure, it was a 'vitalist' one. As we have seen, he conceived of vital functions as residing *irreducibly* in particular tissues of the body. Living matter was distinct from ordinary physical matter. But in view of Lawrence's vitalism, the fact that both he and his adversary – who differed so significantly – are united under the 'vitalist' label certainly puts pressure on its historiographical utility. In the 1970s, Edward Benton (1974, pp. 21–23) proposed that, instead of labelling thinkers as 'vitalists', 'mechanists', or whatever, based upon a superficial glance at the opinions they professed in print, we produce a scheme in which 'vitalist' theories are classified according to variance along the 'dimensions' of: epistemological scepticism; the formal character of the explanations they propose; the fields of study in which different sorts of vitalist explanations were proposed, etc.

More than forty years on, and despite some efforts in this direction, we are still without such a scheme.[32] The entry on 'Materialism and Vitalism' in *The Oxford Companion to the History of Modern Science* is symptomatic of our failings, as we hear that '[m]aterialists make the ultimate principles matter and motion; vitalists, the soul or an irreducible life force'.[33] Certainly there is no room for Lawrence in this dichotomous characterisation. Hasok Chang (2012, p. 111) has suggested that, when we are without 'ready-made philosophical concepts through which a given historical episode can be properly understood, the historian needs to craft new abstract philosophical concepts'. The perspective adopted in this chapter has not been productive enough to provide such concepts, but it has certainly underscored our need for them (Benefit a).

The kind of interplay between HS and PS forwarded in this chapter adds to the variety already advocated in the iHPS literature. Arabatzis' exploration of the benefits of integrating PS into historical work, focuses upon issues 'internal' to the process of writing HS in particular cases. My complementary suggestion holds that lessons from PS can be integrated into HS 'externally' to the particularities of any case-study; PS can guide our choice of case-studies, the methodologies we adopt in studying them, and open our eyes to new possibilities concerning the kinds of lessons we stand to learn from such studies. This chapter began with Kinzel's discussion of historiographical pluralism in HS. Guiding her analysis was the assumption that pluralism in our historical accounts of scientific episodes is an obstacle to be overcome – eradicated so far as is possible – in the pursuit of (the one) historical truth. By reflecting on the lessons of the perspectivism movement in PS, and applying those lessons to a historical case-study, I have contended that historiographical pluralism is – far from something we simply 'have to live with'[34] – something we should actively cultivate.

Notes

1 Some, though, question whether any philosophical lessons can be drawn from historical case-studies. For discussion of these issues, see: Joseph C. Pitt (2001); Jutta Schickore (2011); cf. Katherina Kinzel (2015).
2 Katherina Kinzel (2016), p. 146.
3 Stephen Kellert et al., (2006); Hasok Chang (2015).
4 Ronald Giere is the main advocate of perspectivism. See Giere (2006b).
5 Ronald Giere (2006a).
6 Kinzel's examples include Harry Collins and Allan Franklin's alternative accounts of efforts to detect gravitational waves, and Alan Musgrave and Hasok Chang's differing reconstructions of the Chemical Revolution; Kinzel (2016), Section 7.3.

7 Relatedly, Hasok Chang (2015, p. 380) has urged that '[t]here is much to be gained from a pluralist retelling of even those historical episodes that are widely considered to have been "done to death" already'. His active pluralism consists in writing careful histories of the 'losers' in scientific controversies rather than simply celebrating the winners, and combatting the focus upon 'consensus points and explanations of closure'. To the extent that we can view these as alternative 'perspectives', Chang's programme can be easily accommodated within my own. I owe to Chang the 'done to death' phrase in the title of this chapter.

8 John Abernethy (1814); on John Hunter's influence in the period, see Stephen Jacyna (1983a); on teaching practices in this period, see Pauline Mazumdar (1987).

9 On ether-theories, see Geoffrey N. Cantor and M. J. S. Hodge (1981)

10 Stephen Jacyna (1990).

11 This latter work has been analysed in some detail, with a particular focus on Lawrence's views on biological inheritance; Kentwood D. Wells (1971).

12 George D'Oyley (1820), pp. 6, 5. The accusation of 'materialism' seems misplaced, given that Lawrence's Bichatian approach was in truth a variety of vitalism, as I discuss in section 11.4.

13 Philostratus (1823), p. 116.

14 For example, Sharon Ruston (2005), p. 2.

15 One could of course pursue a multiplicity of other perspectives. Another more 'philosophical' option would be to conceive the affair as, at base, a dispute over *scientific explanation*; in this case, how best to explain the relationship between organised bodies and the vital phenomena they exhibit. It may turn out that our current philosophical concepts of scientific explanation can capture Abernethy and Lawrence's disagreement in illuminating ways (which fall outside of the aims or possibilities of the social-historical perspective summarised above). If not, iHPS scholars might follow Hasok Chang's (2012, p. 111) suggestion that, when we are without 'ready-made philosophical concepts through which a given historical episode can be properly understood, the historian needs to craft new abstract philosophical concepts'.

16 For criticism of Cunningham's framework, see Carin Berkowitz's (2011) review of his *The Anatomist Anatomis'd*.

17 Cunningham (2010), p. 157.

18 Ibid., p. 372.

19 Ibid., p. 371.

20 Abernethy (1817), p. 5.

21 William Bynum (1973), p. 445.

22 Lawrence (1819b), p. 1.

23 Ibid., p. 8.

24 Lawrence (1816), p. 106.

25 Lawrence's view of causation was inspired by the Scottish common-sense school, particularly Thomas Brown's (1818) *Inquiry into the Relation of Cause and Effect*; Lawrence's debt to Brown is evinced by a lengthy footnote on p. 78 of his *Lectures* (Lawrence 1819a).

26 Lawrence (1816), p. 149.

27 Richard Olson (1975), p. 96.

28 On vital spirits as heuristic principles, see: James Larson (1979).

29 Lawrence (1819a), p. 10.

30 Lawrence (1816), p. 121, emphasis added.

31 Ibid., p. 72, emphasis added.

32 See, for example, the work of Charles Wolfe, especially Wolfe (2011, 2017).

33 Wellman (2003).

34 Kinzel (2016), p. 146.

Acknowledgements

My thanks to Jim Secord and Greg Radick for overseeing the various evolutions of this research, to the audience at the Leeds conference in January 2017 for insightful feedback, to Adrian Wilson for literature suggestions, and not least to the editors for their constructive criticism and encouragement.

Bibliography

Abernethy, John, *An Enquiry into the Probability and Rationality of Mr. Hunter's Theory of Life, being the Subject of the first two Anatomical Lectures delivered before the Royal College of Surgeons of London* (London: Longman, Hurst, Rees, Orme and Brown, 1814).

Abernethy, John, *Physiological Lectures, exhibiting a General View of Mr. Hunter's Physiology and of his Researches in Comparative Anatomy, delivered before the Royal College of Surgeons* (London: Longman, Hurst, Rees, Orme and Brown, 1817).

Arabatzis, Theodore, 'What's in It for the Historian of Science? Reflections on the Value of Philosophy of Science for History of Science', *International Studies in the Philosophy of Science*, 31. 1(2017), 69–82.

Benton, Edward, 'Vitalism in Nineteenth-Century Scientific Thought: A Typology and Reassessment', *Studies in History and Philosophy of Science Part A*, 5. 1(1974), 17–48.

Berkowitz, Carin, 'Review of The Anatomist Anatomis'd, by Andrew Cunningham', *The British Journal for the History of Science*, 44. 2(2011), 291–293.

Brown, Thomas, *Inquiry into the Relation of Cause and Effect*, 3rd edn (Edinburgh: Archibald Constable &Co, 1818).

Bynum, William, 'The Anatomical Method, Natural Theology, and the Functions of the Brain', *Isis*, 64. 4(1974), 445–468.

Cantor, Geoffrey N. and M.J.S. Hodge, eds., *Conceptions of Ether: Studies in the History of Ether Theories, 1740–1900* (Cambridge; New York: Cambridge University Press, 1981).

Chang, Hasok, 'Beyond Case-Studies: History as Philosophy', in *Integrating History and Philosophy of Science: Problems and Prospects*, ed. by Seymour Mauskopf and Tad Schmaltz (Dordrecht: Springer, 2012), pp. 109–124.

Chang, Hasok, 'Cultivating Contingency: A Case for Scientific Pluralism', in *Science as It Could Have Been: Discussing the Contingency/Inevitability Problem*, ed. by L. Soler, E. Trizio, and A. Pickering (Pittsburgh: University of Pittsburgh Press, 2015), pp. 359–382.

Christie, J.R.R., 'Ether and the Science of Chemistry: 1740–1790', in *Conceptions of Ether: Studies in the History of Ether Theories, 1740–1900*, ed. by Geoffrey N. Cantor and M.J.S. Hodge (Cambridge; New York: Cambridge University Press, 1981), pp. 85–110.

Cunningham, Andrew, 'The Pen and the Sword: Recovering the Disciplinary Identity of Physiology and Anatomy Before 1800: I: Old Physiology—The Pen', *Studies in History and Philosophy of Biological and Biomedical Sciences*, 33. 4(2002), 631–665.

Cunningham, Andrew, 'The Pen and the Sword: Recovering the Disciplinary Identity of Physiology and Anatomy Before 1800: II: Old Anatomy—The Sword', *Studies in History and Philosophy of Biological and Biomedical Sciences*, 34. 1(2003), 51–76.

Cunningham, Andrew, *The Anatomist Anatomis'd: An Experimental Discipline in Enlightenment Europe* (Farnham: Ashgate, 2010).

Desmond, Adrian, 'Artisan Resistance and Evolution in Britain, 1819–1848,' *Osiris*, 3 (1987), 77–110.

Desmond, Adrian, *The Politics of Evolution: Morphology, Medicine, and Reform in Radical London* (Chicago; London: Chicago University Press, 1989).

D'Oyley, George, 'Abernethy, Lawrence &c. on the Theories of Life', *The Quarterly Review* (July 1819), vol. 22 (London: John Murray, 1820), 1–34.

Figlio, Karl, 'The Metaphor of Organization: An Historiographical Perspective on the Bio-Medical Sciences of the Early Nineteenth Century', *History of Science*, 14 (1976), 17–53.

Fox, Robert, 'Fashioning the Discipline: History of Science in the European Intellectual Tradition', *Minerva*, 44(2006), 410–432.

Giere, Ronald, 'Perspectival Pluralism', in *Scientific Pluralism*, ed. by Stephen Kellert, Helen Longino and Kenneth Waters (Minneapolis; London: University of Minnesota Press, 2006a), pp. 26–41.

Giere, Ronald, *Scientific Perspectivism* (Chicago: University of Chicago Press, 2006b).

Goodfield-Toulmin, June, 'Blasphemy and Biology', *The Rockefeller University Review* (1966), 9–18.

Goodfield-Toulmin, June, 'Some Aspects of English physiology: 1780–1840', *Journal of the History of Biology*, 2. 2(1969), 283–320.

Hull, David, 'Testing Philosophical Claims about Science', In *PSA: Proceedings of the Biennial Meeting of the Philosophy of Science Association* Vol. 2 (Chicago: University of Chicago Press, 1992), pp. 468–475.

Jacyna, L. Stephen, 'Images of John Hunter in the Nineteenth Century', *History of Science*, 21. 1(1983a), 85–108.

Jacyna, L., 'Immanence or Transcendence: Theories of Life and Organization in Britain, 1790–1835', *Isis*, 74. 3(1983b), 311–329.

Jacyna, L., 'Romantic Thought and the Origins of Cell Theory', in *Romanticism and the Sciences*, ed. by Nicholas Jardine and Andrew Cunningham (Cambridge: Cambridge University Press, 1990), pp. 161–168.

Kellert, Stephen, Helen Longino and Kenneth Waters, eds., *Scientific Pluralism* (Minneapolis; London: University of Minnesota Press, 2006).

Kinzel, Katherina, 'Narrative and Evidence: How can Case Studies from the History of Science Support Claims in the Philosophy of Science?', *Studies in History and Philosophy of Science Part A*, 49(2015), 48–57.

Kinzel, Katherina, 'Pluralism in Historiography: A Case Study of Case Studies', in *The Philosophy of Historical Case Studies*, ed. by T. Sauer and R. Scholl (Dordrecht: Springer, 2016), pp. 123–149.

Larson, James L., 'Vital Forces: Regulative Principles or Constitutive Agents? A Strategy in German Physiology, 1786–1802', *Isis*, 70. 2(1979), 235–249.

Lawrence, William, *An Introduction to Comparative Anatomy and Physiology: Being the Two Introductory Lectures Delivered at the Royal College of Surgeons* (London: J. Callow, 1816).

Lawrence, William, *Lectures on Physiology, Zoology, and the Natural History of Man, delivered at the Royal College of Surgeons* (London: J. Callow, 1819a).

Lawrence, William, 'Monsters', in *The Cyclopaedia: Or, Universal Dictionary of Arts, Sciences, and Literature*. Vol. 24, ed. by A. Rees (London: Longman, Hurst, Rees, Orme & Brown, 1819b), pp. 1–16.

Massimi, Michela, 'Perspectivism', in *The Routledge Handbook of Scientific Realism*, ed. by J. Saatsi (London; New York: Routledge, 2018), pp. 164–175.

Mazumdar, Pauline, 'Anatomy, Physiology, and Surgery: Physiology Teaching in Early Nineteenth-Century London', *Canadian Bulletin of Medical History*, 4. 2(1987), 119–143.

Morus, Iwan Rhys, *Shocking Bodies: Life, Death and Electricity in Victorian Britain* (Stroud: The History Press, 2011).

Olson, Richard, *Scottish Philosophy and British Physics, 1750–1880: A Study in the Foundations of the Victorian Scientific Style* (Princeton, N.J.: Princeton University Press, 1975).

Philostratus (Foster of Chelmsford), *Somatopsychonoologia, Showing that Body, Life and Mind Considered as Distinct Essences cannot be Deduced from Physiology* (London: R. Hunter, 1823).

Pitt, Joseph C., 'The Dilemma of Case-studies: Toward a Hereclitian Philosophy of Science', *Perspectives on Science*, 9. 4(2001), 373–382.

Ruston, Sharon, *Shelley and Vitality* (Basingstoke: Palgrave-Macmillan, 2005).

Schickore, Jutta, 'More Thoughts on HPS: Another 20 Years Later', *Perspectives on Science*, 19. 4(2011), 453–481.

Temkin, Owsei, 'Basic Science, Medicine, and the Romantic Era', *Bulletin of the History of Medicine*, 37(1963), 97–129.

Wellman, Kathleen, 'Materialism and Vitalism', in *The Oxford Companion to the History of Modern Science*, ed. by J. Heilbron (Oxford: Oxford University Press, 2003), pp. 490–491.

Wells, Kentwood D., 'Sir William Lawrence (1783–1867): A Study of Pre-Darwinian Ideas on Heredity and Variation', *Journal of the History of Biology*, 4. 2(1971), 319–361.

Wolfe, Charles T., 'From Substantival to Functional Vitalism and Beyond: Animas, Organisms and Attitudes', *Eidos*, 14(2011), 212–235.

Wolfe, Charles T., 'Varieties of Vital Materialism', in *The New Politics of Materialism: History, Philosophy, Science*, ed. by S. Ellenzweig and J.H. Zammito (London; New York: Routledge, 2017), pp. 44–65.

12 Between realism and constructivism

A sketch of pluralism for science education

Wonyong Park and Jinwoong Song

Introduction: 'Making use' of the History and Philosophy of Science

The marriage of the History and Philosophy of Science has resulted in many mutual benefits for both fields, as the other chapters in this volume exemplify. Yet, we may further ask: can the benefits of the integration be 'exported' to some relevant disciplines or practices outside of history and philosophy? This chapter makes a case for using ideas developed by iHPS scholars to examine some implications for a closely related area, namely science education. Throughout the chapter, we intend 'science education' to refer to the teaching of scientific subjects (such as general science, physics, chemistry, and biology) that occurs in primary and secondary education and the early university years – as opposed to the preparation of future science professionals – since our arguments are mostly concerned with raising a democratic citizen in our time. In this sense, science education is where iHPS can be 'put into action', in that iHPS can provide educators some useful ideas for dealing with practical issues in science teaching. In fact, there has been a long tradition in science education that focuses on either using historical and philosophical ideas to better teach scientific concepts, or to solve conceptual issues in the science curriculum, teaching, and learning.[1] As a part of this long-established tradition in education, and as an exploration of the practical utility of iHPS, in this chapter we present a re-examination of the epistemological basis of science education, using the ideas of one very actively discussed topic among iHPS scholars, namely scientific pluralism.

Broadly speaking, scientific pluralism is the idea that science, both in principle and in practice, does not consist of a coherent set of shared methods and principles; instead, it understands science as consisting of multiple groups of scientists embedded in heterogeneous languages, cultures, and practices that interact in numerous ways, but none of which can be easily reduced to the others.[2] When looking at how science is practised today, it would be naïve to say that particle physicists, palaeontologists, and molecular biologists work based on the same sets of aims, methods, or principles, even though no one would deny that they all are scientists. If it is indeed the case that science is essentially plural and disunified, the ramifications for science education would be significant. Indeed, much of what should be taught about science and scientific method (and their nature) and how this should be done is dependent upon our understanding of scientific practice. Since the end of the 1970s science educators have put much effort into examining what the 'nature of science' is and what constitutes scientific practice, however, 'how differently' scientists do different sciences in practice and what that implies in terms of science instruction have drawn surprisingly little attention.

In this chapter we bring together ideas from history, philosophy, sociology, science, and education to address a foundational issue in contemporary science education, namely the realism–constructivism controversy. Our primary task is to advocate for a realist form of pluralism as a moderate alternative to the realist and constructivist extremes in science education, and, in order to do that, we reconstruct ideas of pluralism in a way that reveals its relevance to education. To this end, the chapter is structured as follows. In section 12.2, we begin by reviewing the unsolved debate between realism and constructivism in science education and argue that the realist aims of science are not only important in characterising science itself, but also in characterising science education. In section 12.3, we examine the arguments for pluralism proposed by education theorists and philosophers of science and consider pluralistic realism as one possible way out of the dilemma posed between the realist and constructivist extremes. In section 12.4, three distinct dimensions of pluralism are identified and its importance in school science curricula is discussed. Finally, in section 12.5, we present more practical implications of pluralism in science education by suggesting five potential characteristics of what might be considered 'pluralist' science education.

12.1 Realism versus constructivism in science education

Both in science studies and science education, constructivism is one of the most influential ideas that has prevailed from the second half of the twentieth century onward.[3] Though the usage of 'constructivism' is not univocal in both disciplines, one of the notions the two disciplines share is that 'scientific knowledge originates in the social world rather than the natural world'; thus, the latter plays little role in knowledge construction.[4] Harry Collins's (1974) study on TEA lasers, Bruno Latour's laboratory ethnography (1979, Latour and Woolgar 2013), and Andrew Pickering's (1992) study of the social construction of the quark concept are some landmark studies that contributed to the turn to social constructivism in science studies. These 'sociologists of scientific knowledge' urged that external factors such as social relations, political ideology, power structure, and group interests override logic, rationality, and theory-evidence relation when scientific knowledge is formed and further evolves. Constructivism in science studies soon precipitated new debates over how we should teach science in schools: As a result, it has become one of the most significant ideas in the history of science education that 'science as public knowledge is not so much a "discovery" as a carefully checked "construction" [...] and that scientists construct theoretical entities (magnetic fields, genes, electron orbitals...) which in turn take on a "reality"', as argued by Rosalind Driver (1988, p. 137).

In their steady-selling book *The Golem: What You Should Know About Science*, Harry Collins and Trevor Pinch (1998, pp. 148–149) described a hypothetical 'constructivist' science classroom in which they illustrated what science education might mean in the constructivist understanding of science. In this story, the teacher lets students discover the boiling point of water using a thermometer by reading the thermometer when the water is steadily boiling. As can be easily expected, no one gets the 'right' answer (100°C) unless they already know it. Children get values such as 105°C and 99.5°C, while some of them even fail to get any stable value for various reasons. Of main interest to Collins and Pinch, however, was not what process of inquiry each student had or how they justified their answers; rather, it was what happened during the last ten minutes of the lesson:

> Ten minutes before the end of the experiment the teacher will gather these scientific results and start the social engineering. Skip had his thermometer in a bubble of superheated steam when he took his reading, Tania had some impurities in her water, Johnny did not allow the beaker to come fully to the boil, Mary's result showed the effect of slightly increased atmospheric pressure above sea-level, Zonker, Brian and Smudger have not yet achieved the status of fully competent research scientists.

At the end of the lesson, each child is under the impression that their experiment has proved that water boils at exactly 100°C. What did students learn from the lesson? A sense of 're-negotiation', say Collins and Pinch (1998, p. 149). The core of the students' learning occurred during the 'last ten minutes' when the teacher pointed out what prevented each student from getting 100°C as the boiling point, and why they must accept that it should be 100°C. But is this – that science is only a matter of consensus-reaching process and the resulting 'negotiated' knowledge – really enough? The story shows a very useful way of teaching the socio-political nature of science; however, what is overlooked by the authors is that students can learn little from the whole process about how this relates to reality and the natural world.

After the emergence of social constructivism about scientific knowledge and its radical interpretations, educators began to recognise that the problem of realism is crucial for science education, too. Whether a mind-independent reality exists itself is not a question that can be answered by educators, but the way in which we respond to this question has essential significance in understanding the aims of and the desirable approaches to science education. Science educator Michael Matthews, for example, emphasises the importance of our commitment to reality in science education, saying that it 'affects the rationale we give ourselves, parents, society and students for initiating children into the activities and conceptual schemes of science; and affects the motivation students have for learning science.'[5] He goes further and asks: 'if gravity waves are our creation, why spend so much time and money looking for them?'[6] This reminds us of an important fact: that the commitment to reality gives the stakeholders of science education a strong reason to teach science to children. Science is worth learning because it is about what is real and what is (approximately) true about the natural world. If what we teach and learn – atoms, electrons, evolution, etc. – is only a matter of construction and negotiation, then we will lose much of the justification for its inclusion in school curricula. Underlying our argument throughout this chapter is a recognition that the realist aims of science are not only crucial for science's sake, but for education's sake as well.

Again, this is not to say that constructivism in science education should be rejected in its entirety. The idea that science is a product of social interaction and is affected by both personal and social factors does not mean that it is worthless to teach it. What we should avoid is taking either extreme. Seeing science only as a matter of social-political process does not help escape the realism-constructivism dilemma, nor does seeing it only as a collection of absolute facts about mind-independent reality and *the* scientific method. Contrary to the teacher in Collins and Pinch's story, it is possible to imagine 'deeply realist' teachers who believe that science is rational and divorced from social context; that we have *the* scientific method that guarantees success; that the theories and models are the fruits of that method; and that textbook science is undoubtedly the best reflection of the 'single, complete, and comprehensive account of the natural world'.[7] It seems apparent that this latter view of science is no less dangerous than Collins and Pinch's example, since it is likely to result in a kind of dogmatic teaching and make science education nothing more than linear transfer of canonical knowledge and skills.

12.2 Finding a middle ground: pluralistic realism

How can we preserve the core ideas of constructivism while avoiding radical relativism, and save the rationality of science in classrooms while avoiding dogmatism? Some proposals have been made to resolve the tension between the two. An early solution to the dilemma was proposed by Israel Scheffler (2000), whose idea was developed in the famous debate with his teacher and Harvard colleague Nelson Goodman. Scheffler describes his position as 'pluralism', since it is a synthesis of the objectivist and constructivist ideas. Like constructivism, Scheffler's pluralism preserves not only the possibility but also the desirability of different and incommensurable symbol systems. Due to its commitment to realism, however, pluralism does not drift toward a general relativism that lacks criteria for assessing and evaluating these systems.[8]

Scheffler's pluralism offers a useful middle ground between objectivist realism and radical constructivism that does not drift toward either extreme. But since Scheffler did not speak much about what pluralism means in the context of natural science, it is necessary here to consider the case studies conducted by philosophers of science, who have been actively discussing the pluralism issue since the 1960s. Sandra Harding suggests that there were at least three historical roots of the pluralism, heterogeneity, and disunity discussion of science. One is the moral (and political) motivation that called for recognition of various cultures and subcultures, while another is the so-called practice turn in history and sociology of science that began to focus on scientific practices in labs and field sites. In addition to these, the socio-political environment that caused changes in philosophers' interpretations of positivism and the 'unity of science' also precipitated the pluralism discourse in science studies.[9] The investigation of past and current scientific practice has revealed that many objects of scientific inquiry require multiple approaches for scientific explanation and investigation.[10] From a series of case studies, historians and philosophers of science have given examples of the plurality that occurs in numerous topics from diverse areas of science, such as in physical laws, galactic models, quantum mechanics, behavioural sciences, and electrochemistry.[11] These results seemed contradictory to what was assumed in the monist notion of science, in which any phenomenon (at least ideally) was believed to have a singular, comprehensive explanation consisting of a set of fundamental principles.

It should be noted that pluralism is not a straightforward philosophical idea.[12] Some authors tend to ground their pluralist position on strong metaphysical assumptions such as the 'dappled' or 'disordered' world, while others choose to remain on an epistemological level.[13] While some take pluralism as descriptive of scientific practice, others prefer to add a normative dimension to pluralism.[14] Even when a normative approach is taken, different authors have advanced different positions about how to understand the inconsistencies among pluralities. There is radical pluralism, which often is hard to distinguish from relativism, modest pluralism (in which plurality is viewed as resolvable or at least tolerable), and a continuum of positions between them.[15] Here, however, we do not try to argue in favour of one specific position or another, nor do we propose a new version of pluralism, as that would clearly exceed the scope of this chapter. Still, to serve our purpose – to present a new look at science education – what we mean by pluralism is limited at least to meanings that preserve some commitment to realism, for the aforementioned educational reasons.

It is not difficult to find in the Philosophy of Science literature such a realist version of pluralism that pertains to reality but at the same time recognises and values the diversity of methods and explanations in science, just as Scheffler did. One such example is Hasok Chang's 'active realist' version of pluralism, where science is understood as a commitment to 'learn[ing] about mind-independent reality in all possible ways.'[16] It also follows from this notion of pluralism that multiple theories are not obstacles to, but indeed opportunities for, deeper learning and thus the progress of science. This naturally leads to the normative idea that multiple and plural approaches should be cultivated and that accepting this pluralism about science will lead to a better future. This 'active normative epistemic pluralism' (as Chang, 2012, p. 268, calls it) becomes a fruitful source that we consider an alternative to the realist and constructivist extremes of science. Pointing to this issue, Chang adds:

> It [active-realist science] is to make hypotheses about unobservable nature, to take the trouble of deriving new predictions about their observable consequences, to create new observational capacities, to ask new questions, to create new concepts, to enhance theoretical connections (by unification or reduction), and so on – and to take all these activities as far as possible. Here we can see realism and 'constructivism' dovetailing in an unexpected way, which helps us make sense of statements of realism such as the following by Ludwig Boltzmann from 1890: 'According to my feeling, the task of theory lies in the construction of an image of the exterior world, which exists only in our mind and should serve to guide us in all our thoughts and all our experiments'.[17]

Ronald Giere and Michela Massimi's 'perspectival realism' similarly acknowledges that different knowledge claims such as models and theories are always generated and justified within a particular perspective.[18] While discussing that there is no 'God's eye' view of nature ever available to us, Massimi emphasises that having multiple perspectives *per se* does not deny mind-independent facts and truths. Rather, truth is said to be '*contextualised* within the limits afforded by rival scientific models or rival historical perspectives'.[19] Alex Aylward's chapter in this volume showed an articulation of this view by revisiting the famous Abernethy–Lawrence debate on the nature of life in a new perspective. According to perspectival pluralism it is possible to say that, for example, the phlogiston-based and oxygen-based explanations of combustion lie within two different perspectives, are justified within each perspective but not the other, and are constrained by each perspective's scope and limit. Yet this does not mean truth is relativised. Both theories are pointed to the same natural phenomena under inquiry, but the resulting explanations of the phenomena are being limited within the boundaries of each theory. Both active realism and perspectival realism acknowledge the value of diverse scientific practice as well as scientists' commitment to reality, and this enables us to consider them as possible candidates based on which science education could be conceived.

Having described what we mean by pluralism in science, let us return to educational concerns. One recent pluralist attempt in science education emerged alongside the emphasis on the so-called nature of science in science curricula. In contrast to the earlier approaches to the nature of science that sought to find a definite list of characteristics that represent science, recent discussion has attended more to the irreducibility and disunity of science. Gürol Irzık and Robert Nola attend to the fact that it is hardly possible to speak of necessary and sufficient conditions for something to be deemed science, since all scientific disciplines are different and cannot be reduced. Based on this notion, they argued for a Wittgensteinian 'family resemblance' conceptualisation of the nature of science for education. This conceptualisation was then further developed and articulated by Sibel Erduran and Zoubeida Dagher.[20] As an essential feature of science, Erduran and Dagher discussed several forms of pluralism (pluralism about values, methods, and explanations) and argued that these must be reflected in curriculum materials and teaching practices. In the following sections, we will examine how each of the three categories of pluralism can be viewed through the lens of science education and what they imply.

12.3 Three kinds of pluralism that matter in science education

Pluralism is relevant to a wide range of important issues in science and science education. Stephen Kellert, Helen Longino, and Kenneth Waters (2006a, p. xi) emphasise that the idea of pluralism in science is not restricted to merely recognising that there exist multiple accounts for some natural phenomena, but has broader implications:

> If the nature of the world is such that important phenomena cannot be completely and comprehensively explained on the basis of a single set of fundamental principles, then the aims, methods, and results of the sciences should not be understood or evaluated in reference to the monist quest for the fundamental grail.

However, not all historico-philosophical discussions about disunity and pluralism are equally important in an educational context. For example, most of the metaphysical claims about the structure of the world – whether it is unique, ordered, ruly, or patchy – seem to have little direct relevance to science education, despite their significance in a philosophical context. Given that scientific knowledge and the methods of scientific inquiry are the two most important learning outcomes in science education, it seems more reasonable to start with the epistemic and methodological aspects of scientific pluralism. As one of the few discussions on pluralism in science education to date, Erduran and Dagher implied in their recent work that there are three distinct categories of pluralism that are particularly relevant to science education: pluralism about aims and values, about methods and methodological rules, and about explanations. In what follows, we build on Erduran and Dagher's account of these three classes of pluralism, in order to develop some ideas about how pluralism might be interpreted in a science education context.

12.3.1 Pluralism about the aims and values of science

That scientists' work based on a multiplicity of aims/values and different interpretations of these aims/values has been evidenced by a number of case studies. For example, Peter Galison (1997) has shown with the example of the argumentative and visual traditions in microphysics how diverse epistemic goals are pursued by different scientific communities even within the same field. Similarly, it has also become a popular idea that diverse epistemic and non-epistemic values (sometimes referred as 'virtues') are intertwined in scientific practice. Hasok Chang (2012, p. 207) has illustrated this well in his study of chemical history, showing that 'scientists hold a multiplicity of epistemic values (ranging from empirical accuracy to theoretical elegance), and they also differ on which values they regard as important and how they translate the values into practice'. Aims and values are at the very core of scientific practice, being involved in making decisions about theory choice, and the application of science for specific use.

This realisation has motivated science educators to bring aims and values into the school curriculum during the past few decades, under several rationales.[21] Their discussions led to a close examination of issues such as what the nature of aims and values in science is, how these aims and values are related to other aspects of science, and how teachers should introduce these topics in classrooms. Frequently exemplified as values in science are such epistemic values as generality, refutability, consistency, accuracy, explanatory power and practical values such as simplicity, truth, mathematical elegance, unification. At the same time, many arguments have been made on pluralist grounds to defend the view that abandoning the value-free ideal does not lead to the denial of rationality and objectivity, which implies that we do not have to fear the loss of these virtues when bringing aims and values to classrooms.

Whereas monism about values assumes that there must be one foundational set of values that guides scientific research, value pluralism notes that the list of values and the relative weight among them are not stable but subject to the research problem and aims.[22] Moreover, value pluralism develops to a normative form when combined with the recognition that heterogeneity of values is not a weakness of science, but rather a fruitful source of its strength and objectivity.[23] The idea that scientists work based on certain aims and values that are not determined but are dependent on the group of scientists they belong to is not only important in itself but also as the foundation of pluralism about methods and explanations.

12.3.2 Pluralism about the methods of science

The basic idea of methodological pluralism is that there is no 'single, universally applicable method invariant throughout the History of Science and the various fields of scientific study', and similarly none of the existing methods can unify the others: in short, that there is no method that is *the* method of science. It instead sees science as 'not characterized by a single invariant method, but by a set of evaluative rules to which scientists appeal in the context of theory appraisal.[24] The rise of methodological pluralism is historically linked to logical positivism, which prevailed the early and mid-twentieth century and posited that diverse scientific methods could be united. Against these logical positivist unifiers, defenders of methodological pluralism have provided arguments in favour of the idea that the methods of contemporary science cannot be reduced or unified into one.[25]

The relevance of methodological pluralism to science education becomes clear when we recall, for example, that school science has long been criticised for its reproduction of 'the myth of scientific method'. Henry Bauer (1994) claims that the popular conception that scientists follow a certain method in practice and that the method explains the success of science are myths and misleading. Though this problem has been acknowledged by educators for quite a while, such a simple-minded and mythical view still seems to be widespread among students. Textbooks have tended to present and describe methods similar to Baconian inductivism, hypothetico-deductivism, or some five- to eight-step algorithms labeled as 'the scientific method'. Among others, Erduran and Dagher (2014) recently disproved this by showing that even what is commonly taught as one method (e.g., observation, experimentation) can be divided into many, based on whether its main purpose is description or hypothesis-test, and how manipulative it is.

It is possible to find consolation in such statements of the US *Next Generation Science Standards* as 'scientists use a variety of methods' which appear to indicate that school science is already well versed in pluralism.[26] But is it really sufficient, to expect students to know what to make of such a statement? Perhaps not. Most people already know that what scientists do is different from one field to another. What then should be the main point of teaching methodological pluralism? A good understanding of the plurality of methods consists of not only knowing that different sciences use different methods, but also how and why they are different. This includes understanding how a variety of aims and values can result in the adoption of different methods that (at least appear to) serve those aims and values best and how the choice of a different method can result in a different explanation. While doing an inquiry activity or being introduced to a theory, students should be given the motivation to reflect on such questions in order to help them come to a deeper understanding of the intricate nature of scientific methods. Much more can be done beyond simply telling students that science is plural or that scientists use many methods.

12.3.3 Pluralism about explanations in science

Explanatory pluralism refers to 'the possibility of having more than one explanation that can explain the same phenomena equally well'.[27] This dimension of pluralism is distinct from the preceding two kinds and has driven various case investigations in many different fields of scientific practice. Explanatory pluralism applies to nearly all types of scientific knowledge, which can be broadly grouped into theories, laws, and models.[28] Despite its significance, however, as Erduran and Dagher justly point out, the concept of 'explanatory pluralism' has been neglected despite its utility from scientific and pedagogical standpoints, and this comes from 'the tacit preference for monistic explanations'.[29] The problem of the plurality of explanations being hidden in science curricula reminds us of a remark made as early as 1949, by

British educator Joseph Schwab. In an essay titled *The Nature of Scientific Knowledge as Related to Liberal Education*, he argued that:

> to employ only one doctrine as a principle will give rise to a biased view of the nature of science, and to teach a single doctrine could be to the student only misleading or confusing ... because no single doctrine is more than a partial statement ... based upon a given set of epistemic or metaphysical presuppositions.[30]

It appears, however, that the situation today is not much better than the teaching of 'a single doctrine' that Schwab lamented. Why is it so hard to conceive of plural explanations in school science? One reason is that the plurality is reduced and trimmed in the process of science being transformed in its teachable form. In biology for instance, Erduran and Dagher remark that the selection and organisation of topics does not typically allow any space for pluralism. Another view may be that plurality is obscure because, unlike frontline science, school science deals with 'old' science, where many controversies were resolved long ago and only the 'winner' survives. However, as many historical case studies show, it is clear that plurality is not limited to contemporary science, nor is it only a matter of controversy and survival. In addition, when we look more carefully into school curricula, we discover that they are full of topics with multiple, competing, complementary explanations. In the physical sciences, we see the action-at-a-distance and field models of electrodynamics; the Newtonian, Lagrangian, and Hamiltonian formulations of classical mechanics; the wave and particle models of light; Arrhenius's, Brønsted-Lowry's and Lewis's models of acids and bases; the valence bond and molecular orbital theories of covalent bond. Turning to biology and the earth sciences, more familiar topics are found: the trichromatic and opponent-process theories of colour perception, the punctuated equilibrium and gradualist theories of evolution, the uniformitarianist and catastrophist accounts of diastrophism, a wide variety of models of climate change, etc. Though each case differs from the others in the detailed account of its plurality, all of them are frequently taught in secondary school and at the introductory university level. The reason pluralities are not sufficiently recognised is thus not because they are absent from the curriculum.

The main task for science educators is therefore to make the best use of those pluralistic concepts that are already present in curricula, expose students to them, and teach them how such diversity, plurality, and their co-existence benefit human knowledge about the natural world. By grounding scientific pluralism in our commitment to realism, we can allow students to realise how each of the different explanations reveals a different aspect of reality and how humanity's (and the students') knowledge can be better extended and deepened when these are put together than when we commit to a monistic explanation.

12.4 Re-envisioning science education in light of pluralism: a way forward

Having examined the educational and philosophical dimensions of pluralism, and the possibility of pluralism being compatible with realism, it is time to begin a more systemic reconstruction of these ideas in terms of science teaching and learning in schools. Our task is to recollect the ideas presented thus far and make suggestions for educators to establish curricula and practise pedagogy informed by pluralistic realism. One clear implication of pluralism is that it should teach students that science is a plural discipline composed of a multiplicity of aims, values, methods, and explanations and teach how each one works for the target natural phenomenon. In the following, drawing on the previous discussion about pluralism in science education, we focus on what practical implications pluralism has for science education by suggesting some key characteristics of science education informed by pluralism.

Grounding science education in pluralism, rather than objectivist realism or relativistic constructivism, requires significant re-orientation of what and how we teach in school science. Assembling the core ideas of pluralism articulated by educators and philosophers of science, and transforming it in terms of science education, five broad suggestions can be made for science educators: (a) Science should be taught as a human activity that operates based on multiple aims and values in order to study reality, and not as an established body of knowledge. (b) Science education should encourage students to maximise their close contact with reality by means of pluralism. (c) Students' learning of the scientific method should mean not only acquiring a specific set of methods but also understanding them as a way of learning about reality. (d) Students should learn how to formulate ideas about reality based on shared aims, values, and methods while understanding and mutually recognising others' ideas. (e) Science teachers should encourage students to formulate diverse ideas and approaches while keeping the class directed towards reality. In the rest of this section the meanings of, and rationales for each suggestion are explained in more detail.

(a) Science should be taught as a human activity that operates based on multiple aims and values in order to study reality and not as an established body of knowledge.

The implementation of pluralism in science education can begin by seeing science as a collective human activity that allows multiple aims, values, methods, and explanations, but all these are based on a commitment to reality. This implies one important re-orientation of our views about what should be taught in science classrooms: from teaching science as knowledge to science as practice. This sharply aligns with the recent trend in science education that views science as consisting of a set of scientific practices.[31] Only with science being understood as a collection of epistemic activities and practices in the teaching context can the aims, values, and methods by which the actors (or practitioners) make their decisions become a central goal of teaching science. Scientific knowledge in the form of theories, laws, and models should be understood as the product that humanity has accumulated about reality in the course of this practice, not as a list of absolute and invariant truths that are separate from human activity.

(b) Science education should encourage students to maximise their close contact with reality by means of pluralism.

Another important consequence is that science education should give priority to the maximisation of students' close contact with reality. The aim of a physics lesson is not, for example, 'learning Newton's theory of prismatic refraction' but instead 'learning about prismatic refraction'. A pluralist science lesson puts the reality of natural objects and phenomena at the very centre of it. The so-called hands-on science activities are one (but not the only) way to this: We can let students see, hear, and feel what is happening in the lesson – say, a prism, chemical substances, a bull's eye, a piece of a mineral specimen, etc. – and collectively try to learn from them in the classroom. This can motivate students to recognise what they are learning science for, by giving the strong feeling that they are learning about natural objects and phenomena (in other words, reality) rather than about the theories themselves. The Newtonian, quantum-mechanical, and relativistic views of motion, rather than being taught as three different domains of knowledge isolated from one another, should be appreciated as different lenses through which we approach reality and the truth.

This notion of science learning resonates with Chang's (2012, p. 221) insight that 'for learning to take place, we need to arrange such situations in a way that exposes our senses to the happenings'. Though reality is at the centre of scientific inquiry in such an approach, the importance of theories, laws, and models in science learning is not undermined. Theories, laws and models are used (and understood) as multiple ways to shed light on specific aspects of reality, and when joined together, they allow students to know reality from diverse angles and in a more complete picture. We believe Ronald Giere (2006b, chap. 3) exquisitely demonstrated this benefit of pluralism when he showed, making an analogy with scientific observation, that having a multiplicity of instruments (standard photography, infrared observatory, Compton gamma ray observatory, etc.) can maximise our knowledge of the Milky Way. It is not difficult to infer from Giere's story that while the main aim of science teaching is not mastering individual observational techniques and their background theories, they still can cooperatively serve our purpose – extending knowledge about the outer universe – well. In other words, what science education aims for should not be a monistic truth about reality but rather pluralistic truths about what is in front of us, which is much broader and deeper.

(c) Students' learning of scientific method should mean not only acquiring a specific set of methods, but also understanding them as a way of learning about reality.

In the pluralistic understanding of science, the importance of teaching the methods of science is still recognised. However, scientific methods are not taught as a single stepwise algorithm or a finite list of some kind. Instead, the plurality of methods should be presented as the various forms of human endeavours to learn from reality. Let us give a familiar example; in school physics textbooks the atomic structure is usually explained in accordance with abstract laws of physics, and the light emitted from the shift in atomic state is treated as being an important part of this. On the other hand, in school chemistry, they teach the same object but focus instead on how the properties of atomic structure explain certain chemical bonds and reactions. Both subjects suppose that there is something 'real' about atoms, but the angles of approach differ. This is because each subject has different aims and interests that lead to different methods, and they produce different explanations about reality that serve each of their aims and values. That each account serves its own purposes, however, does not disturb the coherence of knowledge. As a result of this type of process, children learn different truths about the same component of the natural world, while expanding their knowledge and appreciating the values of each method that produced specific explanation about it.

At this level, embracing methodological pluralism in school science does not reduce to a simple-minded idea such as 'scientists use a variety of methods'; it has broader implication for teaching practice. Understanding that different aims/values can produce a diverse range of methods can open up the possibility for students to develop autonomy in judging 'how to do this' (methods) based on 'what I want to do it about' (aims) and 'what is important while doing it' (values). At times students can try different aims and values, which accordingly produce different methods, test them, and maximise their contact with reality. Making this possible from a young age would allow greater opportunities to cultivate the creativity and imagination of individuals.

(d) Students should learn how to make acceptable ideas about reality based on shared aims, values, and methods while understanding and mutually recognising others' ideas.

Through science education informed by pluralism, students should learn how to formulate acceptable ideas about reality at hand, based on shared aims, values, and methods while understanding and recognizing others' ideas. This is particularly important considering that modern society is full of socio-scientific issues, and that living in society as a citizen has required more and more scientific judgement from individuals. That is, most societies require every member to become a scientist to some extent. A citizen in a democratic society is expected to be scientifically literate to make informed decisions about how they evaluate scientific information from the media, whether they vote for a politician who advocates nuclear power plants, GMOs, or climate change caused by humans. This seems to have been the main concern of the developers of *Science for All Americans* in 1991, when they wrote:

Scientific habits of mind can help people in every walk of life to deal sensibly with problems that often involve evidence, quantitative considerations, logical arguments, and uncertainty; without the ability to think critically and independently, citizens are easy prey to dogmatists, flimflam artists, and purveyors of simple solutions to complex problems.[32]

In other words, society requires that students' roles, as democratic citizens, be shifted from acquiring scientific knowledge to making reasoned scientific decisions and judgements. Students agreeing with a particular account of a scientific issue based on the specific values and methods associated with it should be responsible for their own decisions by having an idea of what is occurring inside them during the process of these decisions. In addition, respecting other people's judgement is also an important ability, as long as that judgement is made based on a commitment to reality and is a rational decision based on it. While communicating with their fellow citizens, students should be able to notice the differences between their ideas and also improve their own through interaction.[33] This can be a first step in reducing the destructive form of scientific debate in society and moulding it into collective processes of rational deliberation.

(e) Science teachers should encourage students' diverse ideas and approaches while keeping the class pointed toward reality.

The last implication of pluralism involves the teacher's role; science teachers should motivate students' contact with reality rather than simply transmitting or interpreting established knowledge to students. The existence of a special person called 'teacher' is what distinguishes the science classroom from the scientific community. Professional scientists do not have teachers; they are equal in the sense that all of them participate in the production as well as the evaluation process of knowledge. A science classroom is not identical to a scientific community. The teacher's role in a science classroom is to promote multiple ideas and approaches while letting the multiplicity of students' activities 'point to reality'. That is to say, allowing plurality does not mean rejecting the realist aim of science as the pursuit of truth or facts, nor does it mean speaking against the facts that are valid across perspectives. Plural ideas, arguments, and perspectives should converge to the truth about the subject they are studying on that day (or during the semester).

If we allow plural aims, values, methods, and explanations, some may ask, how can a teacher possibly manage to bring the class together to point to reality? This difficulty can be reduced by recalling that we always base our inquiry on (a set of) aims and values that make the whole process of inquiry, whether in scientific practice or in classrooms. Cohesiveness, adequacy, simplicity, and other values – even if they are not spoken explicitly – guide the classroom as a community that is studying reality, so they should function as valid markers of whether someone's idea is pointing to reality or not. These shared aims and values can sometimes be determined by the subject or topic of each lesson, at other times be based on the teacher's personal epistemology or come from the curriculum goals. The classroom can become a place of rational deliberation in this way.

Therefore, a teacher should be equipped with an ability to guide the classroom to pursue its aims based on shared values, evaluate each student's opinion, sum up, and show directions to follow. This is why we need a teacher in a classroom: because scientists are expected to do all these things by themselves, but in general students are not. The purpose of science lessons being taken this way – that is, as cultivating democratic citizenship through science – teachers should be cautious of their students being obedient to the authority imposed by curricula, text-books, majority opinions, and particularly teachers' own position.

Conclusion: Expanding the toolbox of science education

In this chapter, we sought to explore what the idea of scientific pluralism in science means to science education. We started with the long-standing debate in education between realism and constructivism, proposing that recent forms of pluralism with a commitment to reality can be considered a new epistemic underpinning for understanding science education. To highlight the relevance of pluralism to science education more concretely, three kinds of pluralism (aims/values, methodologies, and explanations) were differentiated, and what each of them means to education was examined. We then proceeded to illustrate what 'pluralist' science education would look like and what kinds of characteristics it should have. We hope to be read as providing an example of iHPS's potential to impact research and practice outside HS and PS themselves.

The disunity, plurality, and diversity of science is not a completely new idea, even in science education. We do not claim originality for bringing pluralism to science education. However, we believe that discussions about pluralism such as the one provided here give a meaningful perspective to understanding the tension between realism and constructivism. We argue that taking pluralism as an approach to science, for policymakers, teachers, and students alike, will allow us to keep our commitment to reality so that we do not lose sight of the reasons why science is being taught (the realist component); simultaneously, it allows multiple ways of inquiry that are grounded on shared goals and epistemic values (the constructivist component), thus giving an impetus to maximise our learning from reality.

We would like to conclude our discussion by addressing a couple of important concerns people may express toward the pluralist approach to science education. The first question that can be raised about our idea is whether it would increase the amount of disciplinary knowledge necessary and whether schools, teachers, and children would be able to handle this increase. This kind of fear can occur when advocating a position that views the pluralist approach to science as 'broadening' our knowledge about nature, a change that would require the broadening of what should be taught in school science while retaining the same number of daily school hours. However, we should note that the growth of human knowledge does not necessarily mean the quantitative expansion of the school curriculum.

Instead of taking pluralism as an addition to knowledge to be transmitted, we can take it as an opportunity to choose from more options and recognise what it means to have a bigger toolbox. Such a notion of pluralism in science education would be crucial in making science learning more purposeful and closer to the practice of science.

A more perplexing issue concerning pluralist science education is the worry that school science might become a place where all students cry out their 'theories' without the teachers having any good reason to reject them. Every day we see people who reject established scientific theories such as evolution and climate change. The widespread 'disrespect for facts', as Harding (2015, p. 111) puts it, has become a serious political issue around the globe. Would pluralism in science education not result in justifying the sceptical attitude toward these facts, eventually preventing any agreement or decisions based on them? Would it not raise more climate change deniers and anti-vaccinators? In order to advocate pluralism in science education, it is important to have a clear answer for these questions.

Compared to a classroom at its realist and objectivist extreme, where a teacher is supposed to become a source of knowledge about reality and transmit it to children in a didactic way, a pluralist classroom surely would look more disorderly and harder to control. However, it is also different from Collins and Pinch's constructivist classroom, which is likely to become a place of 'anything goes'.[34] To borrow Chang's expression, we can instead let 'many things go'.[35] What contemporary pluralists have tried to demonstrate is how such a plural state becomes possible without causing serious problems or inconsistencies, and how it makes science better. We argue that a science classroom can become a space where different aims, values, methods, and explanations come together and where the members expand their knowledge about the subject matter, but all these are rooted in a firm commitment to reality. Understanding pluralism in this way allows us to avoid science education becoming 'transmission' and 'negotiation' and overcome the perils of both realism and constructivism in science education. Furthermore, it offers us important clues for how to equip students for the 'post-truth' world full of 'alternative facts' and 'fake news' delivered via social media. For openness to diversity and critical attitude are not only two key theses of pluralist-realist education, but also are crucial qualities of democratic citizenship in this century. In this sense, we take the 'post-truth' political culture as an opportunity to cultivate rationality and citizenship, rather than a challenge. The ability to explore and appreciate diverse possible sciences and critically assess rival approaches and explanations, we believe, can be best learned in a pluralist science classroom.

Notes

1 One recent contribution to the study of HPS-informed science education is: Michael R. Matthews (2014a).
2 For an introduction to pluralism in philosophy of science, see Alex Aylward's chapter in this volume; Stephen H. Kellert, et al., (2006b).
3 For science studies, a useful reference is Kukla (2013); for science education, see Michael R. Matthews (1998).

4 Steve Woolgar (1983).
5 Michael R. Matthews (1994), p. 158. For a comprehensive discussion about (and sharp attack on) the social constructivism in science education, see e.g. Peter Slezak (1994).
6 Michael R. Matthews (2014b), p. 313.
7 Stephen H. Kellert et al. (2006a), p. x.
8 Katariina Holma (2004), p. 420.
9 Sandra Harding (2015), chap. 5.
10 Stephen H. Kellert et al. (2006b).
11 See Nancy Cartwright (1983) (physical laws); Stephanie Ruphy (2016) (galactic models); Michael Dickson (2006) (quantum mechanics); Helen E. Longino (2002) (behavioural sciences).
12 Jordi Cat (2017). As Cat explains, such meta-pluralism forms an important basis of Kellert, Longino and Waters's so-called *pluralist stance*. See Kellert et al. (2006a).
13 The former view includes John Dupré (1995); Nancy Cartwright (1999). Examples of the latter view are Hasok Chang (2012); Ruphy (2016).
14 Examples of descriptive pluralism includes: Carla Fehr (2006), pp. 167–89; Longino (2002). A form of normative pluralism is found in Chang (2012).
15 Alan Richardson (2006).
16 Chang (2012), chaps 4, 5. John Dupré's (1990) 'promiscuous realism' is another influential realist version of pluralism.
17 Chang (2017), p. 185.
18 Ronald N. Giere (2006a); Michela Massimi (2017).
19 Massimi (2017), p. 171.
20 Richard Rorty (1988); Gürol Irzik and Robert Nola (2011).
21 Douglas Allchin (1999).
22 Longino (2002).
23 Helen E. Longino (1995); Allchin (1999).
24 Howard Sankey (2000), p. 211.
25 Ian Hacking (1996); Dupré (1995); Alison Wylie (1999).
26 NGSS Lead States (2013).
27 Erduran and Dagher (2014), p. 128.
28 Erduran and Dagher (2014), chap. 6.
29 Erduran and Dagher (2014), p. 127.
30 Joseph J. Schwab (1949), p. 99.
31 See, for example, Ellice A. Forman and Michael J. Ford (2014); Cyrus C. M. Mody (2015).
32 American Association for the Advancement of Science (1990), p. xiv.
33 This idea aligns well with Hasok Chang's 'benefit of interaction'. Chang argues that it is also important to note that the co-existence of multiple systems facilitates productive interactions between them through what he calls 'integration, co-optation, and competition' (Chang 2012, chap. 5); It is also to be noted that Jeroen Van Bouwel (2015) further develops this idea into a democratic theory of scientific pluralism.
34 Chang (2012), p. 261.
35 Chang (2012), p. 261.

Acknowledgements

We are grateful to the editors for giving an opportunity to contribute to this volume and for their constructive recommendations. Conversations with Sibel Erduran and Jörg Ramseger greatly helped in sharpening and strengthening the arguments. This work was supported by National Research Foundation of Korea funded by the Ministry of Education (NRF-2016S1A3A2925401).

Bibliography

Allchin, Douglas, 'Values in Science: An Educational Perspective', *Science & Education*, 8(1999), 1–12.

American Association for the Advancement of Science, *Science for All Americans* (Oxford: Oxford University Press, 1990).

Bauer, Henry H., *Scientific Literacy and the Myth of the Scientific Method* (Champaign: University of Illinois Press, 1994).

Cartwright, Nancy, *How the Laws of Physics Lie* (New York: Clarendon Press, 1983).

Cartwright, Nancy, *The Dappled World: A Study of the Boundaries of Science* (Cambridge: Cambridge University Press, 1999).

Cat, Jordi, 'The Unity of Science', in *The Stanford Encyclopedia of Philosophy*, ed. by Edward N. Zalta, Fall 2017 (Stanford: Metaphysics Research Lab, Stanford University, 2017).

Chang, Hasok, *Is Water H₂O?: Evidence, Realism and Pluralism* (Dordrecht: Springer, 2012).

Chang, Hasok, 'Is Pluralism Compatible with Scientific Realism?', in *The Routledge Handbook of Scientific Realism*, ed. by Juha Saatsi (London: Routledge, 2017), pp. 176–186.

Collins, Harry M., 'The TEA Set: Tacit Knowledge and Scientific Networks', *Science Studies*, 4(1974), 165–185.

Collins, Harry M., and Trevor Pinch, *The Golem: What You Should Know about Science*, 2nd edition (Cambridge: Cambridge University Press, 1998).

Dickson, Michael, 'Plurality and Complementarity in Quantum Dynamics', in *Scientific Pluralism*, ed. by Stephen H. Kellert, Helen E. Longino, and C. Kenneth Waters (Minneapolis: University of Minnesota Press, 2006), pp. 42–63.

Driver, Rosalind, 'Theory into Practice II: A Constructivist Approach to Curriculum Development', in *Development and Dilemmas in Science Education*, ed. by Peter Fensham (London: The Falmer Press, 1988), pp. 133–149.

Dupré, John, 'Scientific Pluralism and the Plurality of the Sciences: Comments on David Hull's Science as a Process', *Philosophical Studies*, 60(1990), 61–76.

Dupré, John, *The Disorder of Things: Metaphysical Foundations of the Disunity of Science* (Massachusetts: Harvard University Press, 1995).

Erduran, Sibel, and Zoubeida R. Dagher, *Reconceptualizing the Nature of Science for Science Education* (Dordrecht: Springer, 2014).

Fehr, Carla, 'Explanations of the Evolution of Sex: A Plurality of Local Mechanisms', in *Scientific Pluralism*, ed. by Stephen H. Kellert, Helen E. Longino, and C. Kenneth Waters (Minneapolis: University of Minnesota Press, 2006), pp. 167–189.

Forman, Ellice A. and Michael J. Ford, 'Authority and Accountability in Light of Disciplinary Practices in Science', *International Journal of Educational Research*, 64 (2014), 199–210.

Galison, Peter, *Image and Logic: A Material Culture of Microphysics* (Chicago: University of Chicago Press, 1997).

Giere, Ronald N., 'Perspectival Pluralism', in *Scientific Pluralism*, ed. by Stephen H. Kellert, Helen E. Longino, and C. Kenneth Waters (Minneapolis: University of Minnesota Press, 2006a), pp. 26–41.

Giere, Ronald N., *Scientific Perspectivism* (Chicago: University of Chicago Press, 2006b).

Hacking, Ian, 'The Disunities of the Sciences', in *The Disunity of Science: Boundaries, Contexts and Power*, ed. by Peter Galison and D. J. Stump (Stanford: Stanford University Press, 1996), pp. 37–74.

Harding, Sandra, *Objectivity and Diversity: Another Logic of Scientific Research* (Chicago: University of Chicago Press, 2015).

Holma, Katariina, 'Plurealism and Education: Israel Scheffler's Synthesis and Its Presumable Educational Implications', *Educational Theory*, 54(2004), 419–430.

Irzik, Gürol, and Robert Nola, 'A Family Resemblance Approach to the Nature of Science for Science Education', *Science & Education*, 20(2011), 591–607.

Kellert, Stephen H., Helen E. Longino, and C. Kenneth Waters, 'Introduction: The Pluralist Stance', in *Scientific Pluralism*, ed. by Stephen H. Kellert, Helen E. Longino, and C. Kenneth Waters (Minneapolis: University of Minnesota Press, 2006a), i–xxiiii.

Kellert, Stephen H., Helen E. Longino, and C. Kenneth Waters, eds, *Scientific Pluralism* (Minneapolis: University of Minnesota Press, 2006b).

Kukla, André, *Social Constructivism and the Philosophy of Science* (London: Routledge, 2013).

Latour, Bruno, and Steve Woolgar, *Laboratory Life: The Construction of Scientific Facts* (Princeton: Princeton University Press, 2013).

Longino, Helen E., 'Gender, Politics, and the Theoretical Virtues', *Synthese*, 104 (1995), 383–397.

Longino, Helen E., *The Fate of Knowledge* (Princeton: Princeton University Press, 2002).

Massimi, Michela, 'Perspectivism', in *The Routledge Handbook of Scientific Realism*, ed. by Juha Saatsi (London: Routledge, 2017), pp. 164–175.

Matthews, Michael R., ed, *Constructivism in Science Education: A Philosophical Examination* (Dordrecht: Springer, 1998).

Matthews, Michael R., *International Handbook of Research in History, Philosophy and Science Teaching* (Dordrecht: Springer, 2014a).

Matthews, Michael R., *Science Teaching: The Contribution of History and Philosophy of Science* (London: Routledge, 2014b).

Matthews, Michael R., *Science Teaching: The Role of History and Philosophy of Science* (London: Routledge, 1994).

Mody, Cyrus C. M., 'Scientific Practice and Science Education', *Science Education*, 99 (2015), 1026–1032.

NGSS Lead States, *Next Generation Science Standards* (Washington DC: National Academy Press, 2013).

Pickering, Andrew, *Science as Practice and Culture* (Chicago: University of Chicago Press, 1992).

Richardson, Alan, 'The Many Unities of Science: Politics, Semantics, and Ontology', in *Scientific Pluralism*, ed. by Stephen H. Kellert, Helen E. Longino, and C. Kenneth Waters (Minneapolis: University of Minnesota Press, 2006), pp. 1–25.

Rorty, Richard, 'Is Natural Science a Natural Kind?', in *Construction and Constraint: The Shaping of Scientific Rationality*, ed. by E. McMullin (Notre Dame: Notre Dame University Press, 1988), pp. 49–74.

Ruphy, Stephanie, *Scientific Pluralism Reconsidered: A New Approach to the (Dis) unity of Science* (Pittsburgh: University of Pittsburgh Press, 2016).

Sankey, Howard, 'Methodological Pluralism, Normative Naturalism and the Realist Aim of Science', *After Popper, Kuhn and Feyerabend: Recent Issues in Theories of Scientific Method*, ed. by R. Nola and H. Sankey (Dordrecht: Springer, 2000), pp. 211–229.

Scheffler, Israel, 'A Plea for Plurealism', *Erkenntnis*, 52. 2(2000), 161–173.

Schwab, Joseph J., 'The Nature of Scientific Knowledge as Related to Liberal Education', *The Journal of General Education*, 3(1949), 245–266.

Slezak, Peter, 'Sociology of Scientific Knowledge and Scientific Education: Part I', *Science and Education*, 3(1994), 265–294.

Van Bouwel, J., 'Towards Democratic Models of Science: Exploring the Case of Scientific Pluralism', *Perspectives on Science*, 23(2015), 149–172.

Woolgar, Steve, 'Irony in the Social Study of Science', in *Science Observed: Perspectives on the Social Study of Science*, ed. by K. Knorr-Cetina and M. Mulkay (London: Sage, 1983), pp. 239–266.

Wylie, Alison, 'Rethinking Unity as a "Working Hypothesis" for Philosophy of Science: How Archaeologists Exploit the Disunities of Science', *Perspectives on Science*, 7 (1999), 293–231.

Index